Gerd Faltings,
Gisbert Wüstholz et al.

Rational Points

Aspects of Mathematics

Edited by Klas Diederich

*A Publication of the Max-Planck-Institut für Mathematik, Bonn

Volumes of the German-language subseries "Aspekte der Mathematik" are listed at the end of the book.

Gerd Faltings,
Gisbert Wüstholz et al.

Rational Points

Seminar Bonn/Wuppertal 1983/84

Third Enlarged Edition

A Publication of the
Max-Planck-Institut für Mathematik, Bonn
Adviser: Friedrich Hirzebruch

vieweg

Prof. Dr. *Gerd Faltings* is full professor at Princeton University, Princeton, New Jersey 08544, USA
Prof. Dr. *Gisbert Wüstholz* is full professor at Swiss Federal Institute of Technology, CH-8092 Zürich

AMS Subject Classification: 10BXX, 14G13, 14K10, 14K15

1st edition 1984
2nd edition 1986
3rd enlarged edition 1992

Ursprünglich erschienen bei Friedr. Vieweg & Sohn Velagsgesellschaft
mbH, Braunschweig/Wiesbaden 1992
Softcover reprint of the hardcover 3rd edition 1992

Vieweg is a subsidiary company of the Bertelsmann Publishing Group International.

Cover design: Wolfgang Nieger, Wiesbaden
Printed on acid-free paper

ISSN 0179-2156

ISBN 978-3-322-80342-9 ISBN 978-3-322-80340-5(eBook)
DOI 10.1007/978-3-322-80340-5

Introduction

This booklet consists of the notes of a seminar conducted by the editors during the Wintersemester 1983/84 in Bonn, at the Max-Planck-Institut für Mathematik. The topic was the proof of the Mordell-conjecture, achieved recently by one of us, as well as some additional results about arithmetic surfaces.

We hope that these notes will be useful for mathematicians interested in arithmetic algebraic geometry. We use Arakelov's point of view, which simplifies a lot the classical theory.

The text follows closely the original proof. For a somewhat different point of view (i.e., French versus German style) the reader may consult the exposés Nr. 616/19 in the Séminaire Bourbaki 1983, by P. Deligne and L. Szpiro. L. Szpiro is also conducting a séminaire in Paris, whose notes should be useful as well.

The book is subdivided into seven chapters. The first two, written by G. Faltings, give some general information about moduli spaces and heights. Their main purpose is to define the modular height of an abelian variety, and prove its main properties. Here we often content ourselves with giving descriptions instead of proofs, because the complete details would require at least two additional volumes.

The chapter III, written by F. Grunewald, deals with p-divisible groups and finite flat group-schemes. Its main topic is the relation between Galois-representations and differentials. After those three technical chapters the conjectures of Tate, Shafarevich and Mordell are shown in chapters IV and V, written by N. Schappacher and G. Wüstholz, respectively. In chapter VI G. Faltings gives some complements, mainly the generalization of the results to finitely generated extensions of $\mathbb{Q}$. Finally the chapter VII, by U. Stuhler, contains an introduction to the theory of arithmetic surfaces. (Arakelov's intersection theory, Riemann-Roch, Hodge index-theorem).

We thank the speakers, all participants, the Max-Planck-Institut in general, and its director, F. Hirzebruch. For the typing our thanks go to Mrs. D. Bauer, K. Deutler, and U. Voss.

Bonn/Wuppertal, Gerd Faltings

May 1984 Gisbert Wüstholz

Introduction

Introduction to the Third Edition

Since this book has appeared eight years ago the field has been very active and innovative. New techniques have been developed and surprising new ideas and connections to other fields have been found. As a result there are now five different approaches to the Mordell and Tate conjecture available. Another highlight is the foundation of the arithmetic intersection theory and the proof of the arithmetic Riemann-Roch theorem. Both of these would fill at least another book to make the results accessible to a larger readership.

After the second edition was sold out it seemed clear that another reprint of the book without mentioning the new developments would be rather unsatisfactory. So we decided to try to give in an expanded appendix a report on the new achievements. This turned out to be a tremendous piece of work which delayed the printing of the third edition by at least one year. We leave it to the reader to decide whether this was worthwhile.

Dr. J. Kramer carefully read the manuscript and S. Winiger carefully typed the manuscript for which we thank both of them.

Princeton/Zürich, Gerd Faltings

January 1992 Gisbert Wüstholz

Introduction to the Third Edition

Contents

Appendix: **NEW DEVELOPMENTS IN DIOPHANTINE AND ARITHMETIC ALGEBRAIC GEOMETRY**

(Gisbert Wüstholz)

I

MODULI SPACES

Gerd Faltings

Contents:

§ 1 Introduction

The purpose of this chapter is to list the necessary basic facts from the theory of moduli spaces and their compactifications. Giving complete proofs would require a book, and therefore we usually only describe what is going on. Precise details may be found in the appropriate books, and this survey might be useful as an introduction to them.

The topics we deal with are

- general properties of moduli spaces, and some examples
- logarithmic singularities
- compactification of the complex moduli-space of abelian varieties.

In the next chapter this will be used to define height-functions for abelian varieties over number-fields. I have profited very much from comments and advice given to me by P. Deligne and O. Gabber.

§ 2 Generalities about Moduli-Spaces

Suppose S is a scheme. We want to represent a contravariant functor

$$F : (\text{Scheme}/S)^{O} \to \text{sets}$$

If this is achieved by $M \to S$, we call M a fine moduli-space for F . Dito if we work with algebraic spaces instead of schemes.

In many important cases fine moduli-spaces do not exist. We define a coarse moduli space as an $M \to S$, such that we have a mapping of contravariant functors

$$\phi : F \to h_M = \text{Hom}_S(?,M)$$

with

a) If $T=\text{Spec}(\;) \to S$ is a mapping, with k an algebraically closed field, then ϕ induces a bijection
$F(T) \xrightarrow{\sim} \text{Hom}_S(T,M)$

b) ϕ is universal for mappings $F \to h_N$, that is, for any $N \to S$: $\text{Hom}_S(M,N) \xrightarrow{\sim} \text{Hom}_S(F,h_N)$
obviously b) uniquely determines M .

There are two methods for constructing moduli-spaces, namely geometric invariant theory and Artin's method. We use the latter, and try to explain the main idea.

Suppose first that we want to construct a fine moduli-space M . For any point x of M , the inclusion $\text{Spec}(k(x)) \to M$

($k(x)$=residue-field in x) defines an element of $F(\mathrm{Spec}(k(x)))$. The completion of the local ring of x in M must be the base of a formal universal deformation of this element. If S is of finite type over a field or an excellent Dedekind domain, and if F is a "functor of finite presentation", we can use Artin's approximation theorem to obtain an algebraic scheme $T \to S$ and a point $y \in T$, with $k(y)=k(x) \to M$ extending to an étale mapping from T to M. We thus obtain an étale covering of M.

If we do not have M in advance, we still can make these constructions, and under suitable hypotheses we obtain étale mappings $h_T \to F$ which cover F. In this way we can construct M as an algebraic space. As we have mentioned before, unfortunately in many interesting cases fine moduli-spaces do not exist. This usually happens if we take for F the functor of isomorphism classes of certain objects, like stable curves or principally polarized abelian varieties, and if these objects have nontrivial automorphisms. We then construct a coarse moduli-space, as follows:

Given one of the objects we want to classify, over $\mathrm{Spec}(k)$ with k an algebraically closed field, the finite automorphism group Γ acts on the versal deformation of this object. We algebraisize (following [A]) and obtain an algebraic scheme T with Γ-action, together with a Γ-invariant object of $F(T)$. The coarse moduli-space M then has an étale covering given by the quotients $\Gamma\backslash T$. Usually the "universal object" in $F(T)$ does not descend to $\Gamma\backslash T$. Thus there exists a family of mappings

$$U_i \xrightarrow{p_i} V_i \xrightarrow{q_i} M$$

with q_i étale, p_i finite and dominant, such that

$$M = \bigcup_i p_i(U_i) \ ,$$

and such that over each U_i there exists a "universal object" $\xi \in F(U_i)$. This means that for any geometric point $\mathrm{Spec}(k) \to U_i$, k algebraically closed, the pullback of ξ in $F(\mathrm{Spec}(k))$ is equal to the image of the geometric point in $\mathrm{Hom}_S(\mathrm{Spec}(k),M) \cong F(\mathrm{Spec}(k))$. We shall have to deal with similar situations in the future, where the p_i are allowed to be proper, and so we make the following definition:

Definition:

Suppose M is a noetherian normal algebraic space. A "covering" of M is any finite family of mappings of algebraic spaces

$$\phi_i : U_i \to M \ .$$

with U_i normal, which can be obtained by the following procedure:

a) If the U_i form an étale covering of M, they form a " covering"

b) If there is only one U_i, and if ϕ_i is proper and dominant, we have a "covering"

c) If $\phi_i : U_i \to M$ and $\psi_{ij} : V_{ij} \to U_i$ are "coverings", the compositions

$$\phi_i \circ \psi_{ij} : V_{ij} \to M$$

form a "covering" .

The notion of "covering" has the following properties:

i) $\bigcup_i \phi_i(U_i) = M$

ii) If R is an excellent Dedekind-domain, K its field of quotients, and

$$\psi : \mathrm{Spec}(R) \to M$$

a mapping, there exists a finite extension L of K , and an open covering in the Zariski-topology $\mathrm{Spec}(S)=\bigcup V_i$ (S=normalization of R in L), such that we have commutative diagrams

$$\begin{array}{ccc} V_i & \longrightarrow & U_i \\ \downarrow & & \Big| \\ \mathrm{Spec}(S) & & \Big\downarrow \phi_i \\ \downarrow & & \\ \mathrm{Spec}(R) & \xrightarrow{\psi} & M \end{array} .$$

These properties are easily shown by induction since they are obvious for "coverings" of the types a) and b) above.

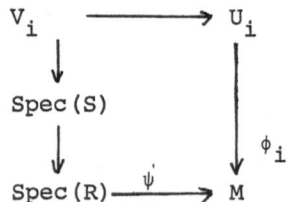

§ 3 Examples

a) Hilbertschemes

Consider a finite type morphism of algebraic spaces $X \to S$, and a finitely presented quasicoherent sheaf $\underline{F}$ on X.

Let for $T \to S$

$$\mathrm{Hilb}_{X/S}(\underline{F})(T) = \left\{ \begin{array}{l} \text{quotients } G \text{ of } F \otimes_{\mathcal{O}_S} \mathcal{O}_T \text{, flat} \\ \text{over } T \text{, whose support is proper}/T \end{array} \right\}$$

Then $\mathrm{Hilb}_{X/S}(\underline{F})$ is representable by an algebraic space locally of finite presentation over S. ([A], Th.6.1). If $X \to S$ is projective and $\mathcal{O}(1)$ an ample line-bundle on X, the space representing $\mathrm{Hilb}_{X/S}(F)$ is the disjoint union of spaces proper over S. Such a decomposition may be obtained via Hilbert-polynomials.

b) Picard-functors

Suppose $f: X \to S$ is finitely presented, proper and flat, and for any $T \to S$ we have $f_*(\mathcal{O}_{X\times_S T}) = \mathcal{O}_T$. Let $\mathrm{Pic}_{X/S}(T)$ be the sheaf in the étale topology associated to $T \rightsquigarrow \mathrm{Pic}(X\times_S T)$. If f has a section $s: S \to X$, we can construct $\mathrm{Pic}_{X/S}(T)$ as

$$\mathrm{Pic}_{X/S}(T) \cong \mathrm{Ker}(s^*: \mathrm{Pic}(X\times_S T) \longrightarrow \mathrm{Pic}(T))$$

Then $\mathrm{Pic}_{X/S}$ can be represented by an algebraic space, locally of finite type ([A],Th. 7.3). We denote it by $\mathrm{Pic}_{X/S}$. We are mainly interested in the case that $f: X \to S$ is a semistable family of curves, that is the geometric fibres are reduced, connected, of dimension 1, and do not contain $\mathbb{P}^1$'s meeting the other components in just one point. We denote by $Pic^0_{X/S} \subseteq Pic_{X/S}$ the subgroup classifying line-bundles whose restrictions to the components of the geometric fibres of f have degree zero. (the corresponding functor can be represented by the same reasons that apply to $Pic_{X/S}$). We have:

Theorem 3.1:

i) $Pic^0_{X/S}$ is separated, smooth, and finitely presented over S.

ii) The fibres of $Pic^0_{X/S} \to S$ are connected, and extensions of abelian varieties by tori.

iii) If f is smooth, $Pic^0_{X/S}$ is proper over S.

Proof:

The statements are local in the étale topology, so we may assume that f has a section

$$s : S \to X .$$

ii) is wellknown. We only indicate that for $S=Spec(k)$, k an algebraically closed field, and $p:\tilde{X} \to X$ the normalization of X, we have an exact sequence

$$\Gamma(X,p_*O^*_{\tilde{X}}/O^*_X) \to Pic^0(X)(k) \to Pic^0(\tilde{X})(k) \to 0 ,$$

where the first term is a product of $(k^\times)$'s, and $Pic^0(\tilde{X})$ an abelian variety.

For iii) we use the valuative criterion, and may assume that S is the spectrum of a discrete valuation ring V, with quotient field K. But then X is regular, and the mapping

$$\mathrm{Pic}(X) \longrightarrow \mathrm{Pic}(X \otimes_V K)$$

is a bijection. (Calculate with divisors. The special fibre is a principal divisor). For i) we first test the separation property with discrete valuation rings. Let V be such a ring, with field of quotients K. We show that the mapping

$$\mathrm{Pic}^{o}(X) \longrightarrow \mathrm{Pic}^{o}(X \otimes_V K)$$

is an injection:

Assume $\underline{L}$ is a line-bundle on X, trivial on the generic fibre. Then $\underline{L} \cong \mathcal{O}(D)$, with a Cartier-divisor D on X whose support is contained in the special fibre. If $C_1 \ldots C_r$ are the irreducible components of the special fibre, D has intersection product zero with each C_j (since it is in Pic^{o}). It is classical that then D is a multiple of the special fibre, and thus a principal Cartier-divisor.

For smoothness we show that for $S=\mathrm{Spec}(A)$ with an artinian ring A, and $I \subseteq A$ an ideal with $I^2=0$, the mapping

$$\mathrm{Pic}(X) \longrightarrow \mathrm{Pic}(X \otimes_A A/I)$$

is a surjection. But its cokernel injects into

$$H^2(X, I\cdot \mathcal{O}_X) = 0 .$$

To show that $\mathrm{Pic}^o_{X/S}$ is finitely presented we may assume that S is noetherian. If $X^o \subset X$ denotes the open subset where f is smooth, we obtain for r big enough a mapping

$$(X^o)^{2r} \twoheadrightarrow \mathrm{Pic}_{X/S},$$

whose image contains $\mathrm{Pic}^o_{X/S}$.

On points this mapping is given by

$$(x_1,\ldots,x_r,y_1,\ldots,y_r) \longrightarrow \mathcal{O}\Big(\sum_{i=1}^{r} x_i - \sum_{j=1}^{r} y_j\Big)$$

Thus $\mathrm{Pic}^o_{X/S}$ is noetherian too.

We also compute the Lie-algebra of $\mathrm{Pic}^o_{X/S}$:

If in general

$$p : G \to S$$

is a smooth algebraic space which is a group, and $s: S \to G$ its zero-section, we let $t^*_{G/S} = s^*(\Omega^1_{G/S})$ and $t_{G/S}$=dual of $t^*_{G/S}$. $t_{G/S}$ and $t^*_{G/S}$ are locally free, and $t_{G/S}$ is called the Lie-algebra of G. It can be determined via deformation theory, and in case that $G = \mathrm{Pic}^o_{X/S}$ with a semi-stable curve $f: X \to S$, we obtain

$$t_{G/S} \cong R^1 f_*(\mathcal{O}_X) ,$$

$$t^*_{G/S} \cong f_*(\omega_{X/S}) ,$$

where $\omega_{X/S}$ denotes the relative dualizing sheaf

c) stable curves

For $g \geq 2$ let

$$\overline{\mathcal{M}}_g(S) = \left\{\begin{matrix}\text{isomorphism classes of stable curves}\\ f : X \to S \quad \text{of genus} \quad g\end{matrix}\right\}$$

there a curve is called stable if it is semistable, and if each smooth $\mathbb{P}^1$ contained in a geometric fibre meets the other components of this fibre in at least three points. $\overline{\mathcal{M}}_g$ has no fine moduli-space, but ([DM]) there exists a coarse moduli-space $\overline{M}_g$, proper over $\mathrm{Spec}(\mathbb{Z})$. This easily leads to

Theorem 3.2:

Suppose S is a noetherian normal algebraic space, $V \subseteq S$ open, and

$$f : X \to V$$

a stable curve. (The genus may vary on the connected components of V, but it is always bigger than one). There exists a "covering"

$$\phi_i : U_i \to S ,$$

and stable curves

$$f_i : X_i \to U_i ,$$

such that over $V_i = \phi_i^{-1}(V)$ X_i is isomorphic to $X \times_V V_i$.

d) principally polarized abelian varieties

Similar to c) we let for $g \geq 1$

$$A_g(S) = \left\{ \begin{array}{l} \text{isomorphism classes of principally polarized} \\ \text{abelian varieties } f:A \to S \text{ , of relative} \\ \text{dimension } g \end{array} \right\}$$

As before there exists a coarse moduli-space A_g over $\mathrm{Spec}(\mathbb{Z})$, but it is not proper. So far we have no reasonable way to compactify it, and this causes a lot of difficulties in the sequel. The method to deal with them is to write an abelian variety as a quotient of a Jacobian (As it was usual in prehistoric times). More precisely, if A/k is an abelian variety over a field k , there exists a smooth complete curve C over k and a surjection

$$\alpha:\mathrm{Pic}^o(C) \to A \ .$$

As $\mathrm{Pic}^o(C)$ is an abelian variety, α has an inverse up to isogeny, that is there exists a $\beta:A \to \mathrm{Pic}^o(C)$ such that $\beta\circ\alpha = d\cdot id$ is multiplication with a natural number $d > 0$.
If k is the generic point of a normal noetherian scheme S , and if A/k is the restriction of an abelian variety A/S , there exists a "covering" $\phi_i:U_i \to S$, such that the pullbacks of C via ϕ_i extend to stable curves C_i over U_i .
Furthermore by the lemma below the pullbacks of α and β can be extended to morphisms

$$\alpha_i : \mathrm{Pic}^o(C_i) \to A\times_S U_i \ ,$$
$$\beta_i : A\times_S U_i \to \mathrm{Pic}^o(C_i) \ , \quad \text{with } \beta_i\circ\alpha_i = d\cdot id.$$

Lemma 3.3

Suppose S is a normal noetherian irreducible algebraic space and A_1 and A_2 semiabelian varieties over S, whose generic fibres are abelian varieties. (The A_i are smooth and separated over S with connected geometric fibres which are extensions of abelian varieties by tori). If $U \subset S$ is a non-empty open set, and

$$\alpha : A_1/U \to A_2/U$$

a morphism over U, α can be extended uniquely to S.

Proof:

The lemma follows from the theory for stable reduction and Néron models if $\dim(S)=1$, especially if S is the spectrum of a discrete valuation-ring. In general we immediately reduce to the case that S is the spectrum of a local ring with algebraically closed residue-field, and that $U=S-\{s\}$, where s denotes the closed point of S. We denote by $Z \subseteq A_1 \times_S A_2$ the closure of the graph of α. We are done if we show that the first projection $pr_1 : Z \to A_1$ is an isomorphism, or that it is proper and injective. (Since A_1 is normal). pr_1 is proper: We use the valuative criterion in the following form:
Let $T = \{t,\eta\}$ be the spectrum of a discrete valuation ring. with t the special and η the generic point. Consider a commutative diagram

$$\begin{array}{ccccc} \{\eta\} & \xrightarrow{\psi} & Z & \xrightarrow{pr_2} & A_2 \\ \downarrow & \nearrow & \downarrow pr_1 & & \\ T & \xrightarrow{\phi} & A_1 & & \\ & & \downarrow f & & \\ & & S & & \end{array}$$

with $f\circ\phi(\eta)\in U$.

We have to show that ψ can be extended to T . It suffices if $pr_2\circ\psi$ can be extended. For this we look at the diagram

$$\begin{array}{ccccc} T & \xrightarrow{\phi\times_S id} & A_1\times_S T & \overset{\tilde{\alpha}}{\dashrightarrow} & A_2\times_S T \\ \uparrow & & \uparrow & & \uparrow \\ \{\eta\} & \longrightarrow & A_1\times_S\{\eta\} & \xrightarrow{\alpha\times_S\{\eta\}} & A_2\times_S\{\eta\} \end{array}$$

If we apply the result for discrete valuation rings as base we obtain an extension $\tilde{\alpha}$ of $\alpha\times_S\{\eta\}$. Then

$$\tilde{\alpha}\circ(\phi\times_S id):T \longrightarrow A_2\times_S T$$

defines a mapping α from T to A_2 , which extends $pr_2\circ\psi$, since the image of ψ lies in the graph of α and hence

$$pr_2\circ\psi = \alpha\circ(\phi|\{\eta\}) .$$

pr_1 is injective:

We show that for any point $X\in A_1\times_S\{s\}$ with $k(x)=k(s)$ there exists at most one point

$$(x,y)\in Z\times_S\{s\}\subseteq(A_1\times_S A_2)\times_S\{s\}$$

$(y\in A_2\times_S\{s\})$

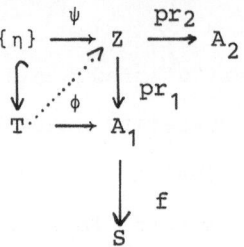

For this we first need some general remarks: If $T=\{\eta,t\}$ is the spectrum of a discrete valuation ring and $\psi:T\to S$ a mapping with $\psi(\eta)\in U, \psi(t)=s$, we can extend α after base-change to an

$$\tilde{\alpha}:A_1 \times_S T \to A_2 \times_S T$$

The induced morphism

$$A_1 \otimes_{k(s)} k(t) \to A_2 \otimes_{k(s)} k(t)$$

is already defined over $k(s)$, since this field is algebraically closed and since $A_1 \otimes_{k(s)} k(t)$ is semiabelian. (Use l-division points!) It is thus induced from an

$$\alpha_s: A_1 \times_S \{s\} \to A_2 \times_S \{s\} \quad .$$

This α_s is independant of the choice of T and ψ, since its effect on l-division points is determined by the map α over U. Now we claim that with our previous notations necessarily $y=\alpha_s(x)$: There exist T as above and $\psi:T\to Z \subseteq A_1 \times_S A_2$ with $f\circ pr_1 \circ \psi(\{\eta\})\in U$, $\psi(t)=(x,y)$. From the commutative diagram

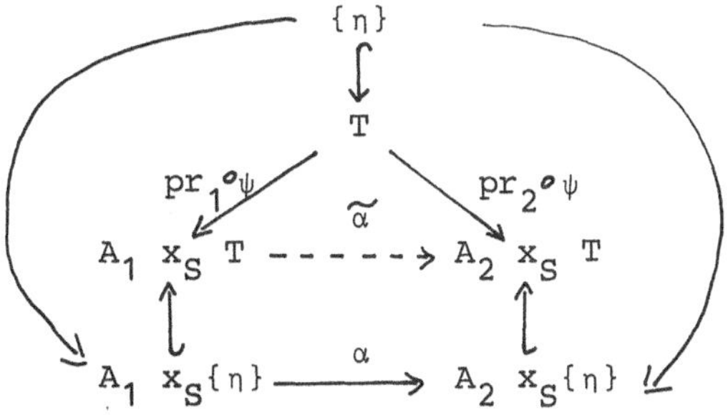

we see that indeed $y=\tilde{\alpha}(x)=\alpha_s(x)$.

§ 4 Metrics with logarithmic singularities

In the sequel we shall need various hermitian metrics on vector bundles, which have mild singularities. To formalize the situation we make the following definition:

Definition

Let X be a normal complex space, $Y \subseteq X$ a closed analytic subset such that $U = X-Y$ is dense in X. If $\underline{E}$ is a vector-bundle on X and $\langle , \rangle$ a hermitian metric on $\underline{E}/U$, this metric has logarithmic singularities along Y if the following holds: For $y \in Y$, there exist a neighbourhood V of y in X, holomorphic functions $f_1,\dots,f_l$ on V with Y as common set of zeroes, and sections $e_1,\dots,e_r$ of $\underline{E}$ over U which form a basis of $\underline{E}/U$, such that for some constants $c_1, c_2 > 0$,

$$|\langle e_i, e_j \rangle|(z) \leq c_1 \cdot |(\log(\max(|f_i(z)|)|^{c_2}$$

$$|\det\langle e_i, e_j \rangle|(z)^{-1} \leq c_1 \cdot |(\log(\max(|f_i(z)|)|^{c_2}$$

for $z \in U \cap V$.

Remarks:

a) The extension $\underline{E}$ of $\underline{E}/U$ is uniquely determined by this property, since a local section of $\underline{E}/U$ is holomorphic on X if and only if its norm grows at most logarithmically near Y.

b) The definition is essentially independant of the choice of the f_i and e_j:
For another choice f_i and $\tilde{e}_j$, and for a neighbourhood $W \subset\subset V$ of y inequalities like the one above hold for the new data.

c) If $<,>_1$ is a hermitian metric on $\underline{E}$ (not only on $\underline{E}|U$), $<,>$ has logarithmic singularities if and only if for any $y \in Y$ we can find $V, f_1, \ldots, f_r, c_1, c_2 > 0$ as above, such that

$$c_1^{-1} \, |\log(\max(|f_i(z)|))|^{-c_2} \cdot \|e\|_1(z)$$
$$\leq \|e\|(z)$$
$$\leq c_1 \cdot |\log(\max(|f_i(z)|))|^{c_2} \cdot \|e\|_1(z) ,$$

for any section e of $\underline{E}$ over $U \cap V$, and $z \in U \cap V$. ($\|e\|$, $\|e\|_1$ are the norms defined by $<,>$, $<,>_1$.)

d) If $\underline{F} \subseteq \underline{E}$ is a subbundle such that $\underline{E}/\underline{F}$ is locally free too, a metric with logarithmic singularities on $\underline{E}$ induces such a metric on $\underline{F}$ and $\underline{E}/\underline{F}$ (Use c))

e) If $\underline{E}$ has a metric with logarithmic singularities, so have $\underline{E}^*$, $S^p(\underline{E})$, $\Lambda^p(\underline{E})$, etc.

f) If (X_1, Y_1) is a pair fulfilling the assumption on X and Y, and $\phi: X_1 \to X$ a holomorphic map with $Y_1 \supseteq \phi^{-1}(Y)$, then the pullback of a hermitian metric on $\underline{E}|U$ with logarithmic singularities along Y is a hermitian metric on $\phi^*(\underline{E})/U_1$ with logarithmic singularities along Y_1. The converse is true (i.e., $\phi^*<,>$ logarithmic singularities $\Rightarrow$ $<,>$ logarithmic singularities) if ϕ is proper and surjective, and $Y_1 = \phi^{-1}(Y)$.

Examples of metrics with logarithmic singularities arise as follows:

Theorem 4.1

Suppose (X,Y) fulfill the assumptions of the definition,

and let

$$f : C \to X$$

be a family of semistable curves, with good reduction outside Y, that is:

i) f is proper and flat

ii) The fibres of f are semistable curves of genus ≥ 2.

iii) For $x \in U$ $f^{-1}(x)$ is a non-singular curve..

Let $\underline{E}=f_*(\omega_{C/X})$. Then $\underline{E}|U \cong f_*(\Omega^1_{C/X})|U$, and square integration of differentials on the fibres defines a hermitian metric on $\underline{E}|U$. This hermitian metric has logarithmic singularities along Y.

<u>Proof:</u>

The claim is local along Y. Choose $y \in Y$, and let $C(y)=f^{-1}(\{y\})$ be the fibre at y. The fibration f is locally induced from a versal deformation of $C(y)$. We may assume that it is the versal deformation, and Y the diskriminant locus.

Denote by g the genus of $C(y)$, by Δ the unit-disk ($|z|<1$), and let $X=\Delta^{3g-3}$, $y=(0,\dots,0)$, be the base of the versal deformation of $C(0)=C(y)$. We assume that $Y \subseteq X$ is the discriminant locus, so that Y is a union of hyperplanes. From the deformation-theory of semistable curves we know that there is an open covering $C = \bigcup_{j=1}^{r} U_j$, such that either $f|U_j$ is smooth, and thus $U_j \cong \Delta \times X$, or that

$$U_j \cong \{(z,w,x) \in \Delta\times\Delta\times X \mid z\cdot w = f_j(x)\},$$

where $f_j(x)$ is the defining equation of one of the components of Y. The mapping f is given by

$$f((z,x)) = x \quad \text{resp.} \quad f((z,w,x))=x \qquad (z,w\in\Delta)$$

The relative dualizing complex is generated over U_j by dz resp. dz/z. Thus a section α of $f_*(\omega_{C/X})$ is given by

$$\alpha = \phi(z,x)dz \quad \text{resp.} \quad \alpha = \phi(z,w,x)\cdot dz/z,$$

with ϕ holomorphic.
To get one of the inequalities necessary for logarithmic singularities we estimate from above

$$\int_{f^{-1}(x)} |\alpha|^2 \leq \sum_j \int_{f^{-1}(x)\cap U_j} |\alpha|^2$$

The integral over the U_j with $f^{-1}(x)\cap U_j$ smooth remains bounded, and we come down to estimating

$$\int_{\substack{z\cdot w=f_j(x)\\ |z|<1\\ |w|<1}} \left|\frac{dz}{z}\right|^2$$

$$= \int_{|f_j(x)|<|z|<1} \left|\frac{dz}{z}\right|^2 = 2\pi \int_{|f_j(x)|<r<1} \frac{dr}{r} = -2\pi\log|f_j(x)|,$$

if $|f_j(x)| \leq 1$.

As the zero-set of f_j is contained in Y, we have one of the necessary inequalities. The other one is quite easy:

We may assume that the U_j with $f|U_j$ smooth meet each irreducible component of each fibre $f^{-1}(x)$. If $<,>_1$ is a hermitian metric on $f_*(\omega_{C/X})$ on X, there is a $c > 0$ with

$$\sum_{f|U_j \text{ smooth}} \int_{f^{-1}(x)\cap U_j} |\alpha|^2 \geq c \cdot \|\alpha\|_1^2$$

(If $\alpha|f^{-1}(x)\cap U_j$ vanishes for each j with $f|U_j$ smooth, α vanishes on $f^{-1}(x)$).

Remark:

The semiabelian group algebraic space $A=\mathrm{Pic}^0(C_i)$ is principally polarized over U. Any polarization induces a hermitian metric on $t^*_{A/X}|U$, and as two polarizations can be compared we see that all such metrics have logarithmic singularities along Y. This result extends to arbitrary semiabelian varieties.

§ 5 The minimal compactification of $A_g/\mathbb{C}$.

We give an analytic description of the moduli-space A_g over the complex numbers, and of its compactification A_g^* which has been constructed by Satake. The construction has been generalized by Baily-Borel [BB], and it has the property that any analytic mapping

$$X^o \to A_g ,$$

with X^o algebraic, can be extended to an algebraic mapping $X \to A_g^*$, where $X \supseteq X^o$ is a compactification.

We first give the analytic description of A_g:
A principally polarized abelian variety $A/\mathbb{C}$ of dimension g can be given by its cohomology $U=H^1(A,\mathbb{Z})$, together with an unimodular sympletic form $\langle , \rangle : U \times U \to \mathbb{Z}$ and a totally isotropic subspace of dimension g

$$V = \Gamma(A,\Omega_A^1) \subseteq U \otimes_{\mathbb{Z}} \mathbb{C} = H^1(A,\mathbb{C}) ,$$

such that

$$i\cdot\langle v,\bar{v}\rangle > 0$$

for $v \in V$, $v \neq 0$.

It is known that the pairs $(U,\langle , \rangle)$ are all isomorphic. So we may assume that $U=\mathbb{Z}^{2g}$ with basis $e_1,\dots,e_g, f_1,\dots,f_g$, and

$$\langle e_i,e_j\rangle = \langle f_i,f_j\rangle = 0 ,$$
$$\langle e_i,f_j\rangle = -\langle f_j,e_i\rangle = \delta_{ij}$$

The automorphisms of $(U, <,>)$ then are equal to $G(\mathbb{Z})$, where $G=Sp(2g)$ denotes the symplectic group. In the sequel we write $U_{\mathbb{Q}}$ for $U \otimes_{\mathbb{Z}} \mathbb{Q}$, and similar $U_{\mathbb{R}}, U_{\mathbb{C}}, G(\mathbb{Q}), G(\mathbb{R}), G(\mathbb{C})$. We define

$$\check{D} = \{\text{totally isotropic complex subspaces } V \subseteq U_{\mathbb{C}} \text{ of dimension } g\}$$

$$D = \{V \in \check{D},\ i<v,\bar{v}> > 0 \text{ for } v \in V, v \neq 0\} \subset \check{D}.$$

$\check{D}$ is a Zariski-closed subset of some Grassmannian, and homogeneous under $G(\mathbb{C})$. D is open in $\check{D}$, and homogeneous under $G(\mathbb{R})$. By the previous considerations we know that

$$A(C) \cong {}_{G(\mathbb{Z})}\backslash^{D} .$$

This is in fact an isomorphism of analytic spaces. If $\Gamma \subset G(\mathbb{Z})$ denotes a neat subgroup of finite index (for example a suitable congruence subgroup) there exists a principally polarized abelian variety A over $X={}_{\Gamma}\backslash^{D}$, whose fibre over the equivalence class of $V \in D$ is given by

$$A = V^*/U^* . \quad (V^*=\mathrm{Hom}_{\mathbb{C}}(V,\mathbb{C}) , U^*=\mathrm{Hom}_{\mathbb{Z}}(U,\mathbb{Z}))$$

The bundle $t^*_{A/X}$ is the bundle defined on X by taking the Γ-quotient of the $G(\mathbb{R})$-equivariant bundle on D given by the V's.

For a real isotropic subspace $W \subseteq U_{\mathbb{R}}$ we define a subset $F(W) \subseteq D$ by $V \in F(W)$ if and only if

i) $i\langle v,\bar{v}\rangle \geq 0$ for $v \in V$

ii) $W_{\mathbb{C}} = \{v \in V \mid i\langle v,\bar{v}\rangle = 0\}$

Thus $F(0)=D$, and the topological closure $\bar{D}$ of D in $\check{D}$ is given by

$$\bar{D} = \bigcup_{W \text{ isotropic}} F(W)$$

$F(W)$ is homogeneous under $Sp(W^{\perp}/W)$ and isomorphic to the object we obtain if we start in the definition of D with $W^{\perp}/W$ instead of U.

Such an $F=F(W)$ is called a boundary component of D, and F is defined to be rational if W can be defined over $\mathbb{Q}$. To simplify notations we write $F(W)=F(W_{\mathbb{R}})$ for an isotropic subspace $W \subseteq U_{\mathbb{Q}}$.

We let

$$D^* = \bigcup_{\substack{W \subseteq U_{\mathbb{Q}} \\ \text{isotropic}}} F(W) .$$

D^* is stable under $G(\mathbb{Q})$. If $\Gamma \subseteq G(\mathbb{Z})$ is a subgroup of finite index, $X^* = {}_{\Gamma}\backslash D^*$ has the structure of a normal compact complex space. If $\Gamma = G(\mathbb{Z})$, then $X^* = A_g^*$ is as a set the disjoint union

$$A_g^* = A_g \cup A_{g-1} \cup \dots\dots \cup A_0 ,$$

where A_j for $0 \leq j \leq g$ corresponds to the quotient

$$\setminus \bigcup_{\dim(W)=g-j} F(W) \quad .$$

(All W of the same dimension are conjugate under $G(\mathbb{Z})$.

It is known that X^* is a projective algebraic variety. An ample line-bundle can be described as follows:
For some $r > 0$, the r'th power of the $G(\mathbb{R})$-equivariant bundle on D defined by the $\Lambda^g V$ gives a line-bundle on $X = {}_\Gamma\backslash^D$. (If Γ is neat, we may take $r=1$, and obtain $\Lambda^g t^*_{A/X}$). This line-bundle extends to X^* , and is ample. The proofs of these facts can be found in [BB] .

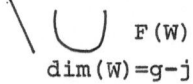

§ 6 The toroidal compactification

We want to construct a non-singular model X^+ of X^*. To avoid the difficulties arising from singularities of $X = {}_\Gamma\backslash^D$ we assume that Γ is neat.

We first consider the realization of D is a Siegel-domain: Let $W \subseteq U_{\mathbb{Q}}$ be an isotropic subspace, $F=F(W)$. $N(F) \subseteq G$ denotes the parabolic subgroup of symplectic transformations which fix W (and hence also $W^\perp$). We choose a Levi-decomposition of $N(F)$, which amounts to choosing an isotropic subspace $W^0 \subseteq U_{\mathbb{Q}}$ such that $U_{\mathbb{Q}} = W^0 \oplus W^\perp$. This leads to an orthogonal decomposition

$$U_{\mathbb{Q}} = (W \oplus W^0) \oplus (W^\perp \cap W^{0\perp}) ,$$

and W and W^0 are dual to each other.

Let

$G_h(F)$ = Stabilizer of $(W \oplus W^0)$ in G
$\cong Sp(W^\perp \cap W^{0\perp}) \cong Sp(W^\perp/W)$,

$G_l(F) = \{ (\alpha, {}^t\alpha, id) \mid \alpha \in Aut(W) \}$

($G_l(F)$ operates trivially on $W^\perp \cap W^{0\perp}$,

$R(F) = \{\alpha \in N(F) \mid \alpha$ operates trivially on $U_{\mathbb{Q}}/W^\perp, W^\perp/W$ and $W\}$

Then $R(F)$ is the unipotent radical of $N(F)$, and

$$N(F) = (G_h(F) \times G_l(F)) \times R(F).$$

is a Levi-deompositon.

Furthermore,

$$R(F) = \exp(\underline{v}(F))\cdot U(F)$$
$$= \exp(\underline{v}(F))\cdot\exp(\underline{u}(F))$$

with

$$\underline{v}(F) = \{C\ ,\ C \in \underline{g}\ ,\ C(W^0) \subseteq W^{\perp} \cap W^{0\perp},\ C(W^{\perp} \cap W^0) \subseteq W,\ C(W) = 0\ \}$$

$$\underline{u}(F) = \left\{H \in \underline{g}\ \middle|\ H(U) \subseteq W,\ H(W^{\perp}) = 0\right\}$$

($\underline{g}$=Lie(G) = $\underline{sp}(2g)$)

$\underline{v}(F)$ is a subspace of $\underline{g}$, and $U(F)$ is the centre of $R(F)$.
Furthermore

$$\underline{u}(F) \cong S^2(W^0)^* = \left\{\begin{matrix}\text{symmetric bilinear forms}\\ \text{on } W^0\end{matrix}\right\}$$

where to $H \in \underline{u}(F)$ corresponds the quadratic form with value $\langle w,H(w)\rangle$ for $w \in W^0$. $C(F) \subseteq \underline{U}(F)_{\mathbb{R}}$ denotes the cone of positive definite quadratic forms, and we frequently identify $U(F)_{\mathbb{R}}$ via exp with $\underline{u}(F)$.

For $V \in F=F(W)$ we have

$$V = W \oplus (V \cap W^{\perp} \cap W^{0\perp})$$

Define

$$V^0 = W^0 \oplus (V \cap W^{\perp} \cap W^{0\perp}) \in F^0 = F(W^0)\ ,$$

and

$$\lambda: F \times \underline{v}(F)_{\mathbb{R}} \times \underline{u}(F)_{\mathbb{C}} \to \check{D}$$

by

$$\lambda(V,C,A+iB) = \exp(A+iB)\cdot\exp(C)\cdot(V^0)\ .$$

Theorem 6.1:

i) $\operatorname{Im}(\lambda) = D_F = \{V \mid i<v,\bar{v}>>0 \text{ for } v \in V \cap W^{\perp}, v \neq 0\}$

ii) λ induces a bijection

$$F \times \underline{v}(F)_{\mathbb{R}} \times \underline{u}(F)_{C} \cong D_F$$

iii) $\lambda^{-1}(D) = F \times \underline{v}(F)_{\mathbb{R}} \times (\underline{u}(F)_{\mathbb{R}} + i\mathbb{C}(F))$

Remark: λ is a diffeomorphism.

Example:

If W is maximal isotropic, then $\underline{v}(F)=0$, F is a point, and

$$D \cong \{X+iY \in M(g,\mathbb{C}) , \; X,Y \text{ symmetric, } Y \text{ positive definite}\}$$

This is the classical Siegel upper half-plane $\mathbb{H}_g$.

Proof of the theorem:

i) Obviously $F^{o}=F(W^{o})$ is contained in D_F , and D_F is stable under $R(F)$ and $\exp(i\underline{u}(F)_{\mathbb{R}})$, since $R(F)$ stabilizes W and $\underline{u}(F)_{\mathbb{R}}$ annihilates $V \cap W^{\perp}$ (so $\exp(iB)(V \cap W^{\perp}) = V \cap W^{\perp}$, for $B \in \underline{u}(F)_{\mathbb{R}}$) Therefore $\operatorname{Im}(\lambda) \subseteq D_F$.

On the other hand $\operatorname{Im}(\lambda)$ is stabilized by the group $N(F)\cdot\exp(\underline{u}_{F,\mathbb{C}}) = N(F)\exp(i\underline{u}_{F,\mathbb{R}})$, and it suffices to show that D_F has only one orbit under this group. As D is homogeneous under $N(F)$, we are done if we show that $D_F = \exp(i\cdot\underline{u}(F)_{\mathbb{R}})D$. As $D \subseteq D_F$, it is clear that the right hand side is contained in the left one, so let us choose $V \in D_F$. We need an element $B \in \underline{u}(F)_{\mathbb{R}}$ with $\exp(i\cdot B)(V) \in D$. This means that for $v \in V$, $v \neq 0$, the hermitian form

$$
\begin{aligned}
& i < \exp(iB)v, \overline{\exp(iB)v} > \\
= & \ i < \exp(iB)v, \exp(-iB)\overline{v} > \\
= & \ i < v, \exp(-2iB)\overline{v} > \\
= & \ i \cdot <v,\overline{v}> + 2<v,B\cdot\overline{v}>
\end{aligned}
$$

takes a positive value.

This is the case if $v \in V \cap W^{\perp}$, since then B annihilates v. On the other hand if $E \subseteq V$ denotes the space of elements perpendicular to $V \cap W^{\perp}$ for the hermitian product above (or for $i<v,\overline{v}>$, which leads to the same E), then E injects into $U_{\mathbb{C}}/W^{\perp} \cong W^{O}_{\mathbb{C}}$, and for $v \in E$ $<v,B\cdot\overline{v}>$ is the value of the hermitian scalar-product defined by the symmetric bilinear form $B \in \underline{u}_{F,\mathbb{R}} \cong S^2(W^O)^*$ on the image of v in $W^O_{\mathbb{C}}$. If we choose B te be sufficiently positive definite, we obtain what we need.

ii) We want to recover $A,B \in \underline{u}(F)_{\mathbb{R}}, C \in \underline{v}(F)_{\mathbb{R}}, V \in F$ from $\exp(A+iB)\cdot\exp(C)(V^O)$. It is easy to find V:

$$V = W \oplus (V \cap W^{O\perp}) = W \oplus (V \cap W^{\perp} \cap W^{O\perp}),$$

and $V \cap W^{\perp} \cap W^O$ has the same image in $W^{\perp}/_W \cong W^{\perp} \cap W^{O\perp}$ as $\exp(A+iB)\exp(C)(V^O) \cap W^{\perp}$.

We thus may fix V and assume that $\exp(A+iB)\exp(C)$ stabilizes V^O. We want to show that $A=B=C=0$.

If

$$v \in V^O \cap W^{\perp} = V \cap W^{O\perp}, \text{ then}$$
$$A(v) = B(v) = 0, \quad (\text{since } V^O \subseteq W^{O,\perp})$$

and

$$\exp(C)(v) - v = C(v) \in W \cap V^O = (0)$$

so C annihilates $V^0 \cap W^{\perp}$. C is real, so it annihilates the complex conjugate $\overline{V^0 \cap W^{\perp}}$, hence also

$$W^{\perp} \cap W^{0\perp} = (V^0 \cap W^{\perp}) \oplus \overline{(V^0 \cap W^{\perp})}$$

(if v and $\bar{v}$ lie in $V^0 \cap W^{\perp}$, $i\langle v,\bar{v}\rangle=0$, so $v=0$).

As C is skew-symmetric for $\langle , \rangle$, and as $C(W^0) \subseteq W^{\perp} \cap W^{0\perp}$

$C(W^0)=0$, *and*

$$C = 0 \quad .$$

If now $\quad v \in W^0_{\mathbb{C}}$,

$$(A+iB)(v) = \exp(A+iB)(v) - v \in V^0 \cap W_{\mathbb{C}} = (0) \quad ,$$

so $A+iB$ annihilates $W^0_{\mathbb{C}}$ and $A=B=0$.

iii) The necessary computations have already been made in i): If $V \in F$, $V^0 = W_{\mathbb{C}}{}^0 \oplus (V \cap W^{0\perp})$ is the orthogonal decomposition for the scalarproduct $i\langle v,\bar{v}\rangle$ used in i), where $W^0_{\mathbb{C}}$ was denoted by E. Note that $i\langle v,\bar{v}\rangle = 0$ for $v \in W^0_{\mathbb{C}}$

Thus:

$$\exp(A+iB)\exp(C)(V^0) \in D$$

$$\Longleftrightarrow \exp(iB)(V^0) \in D$$

$$\Longleftrightarrow \langle v, B\bar{v}\rangle > 0 \quad \text{for} \quad v \in W^0_{\mathbb{C}},\ v \neq 0$$

$$\Longleftrightarrow B \in C(F) \quad .$$

We now give a local description of a smooth compactification X^+ of $X = \Gamma\backslash D$. We remind the reader that Γ is supposed to be neat.

The construction of X' makes use of rational polyhedral decompositions of the cones $C(F)$, for all boundary components F. The details can be found in [AMRT]. To give the idea we make the following construction.

Construction.

Choose a rational boundary component F. Then $\Gamma\cap U(F)(\mathbb{R})$ is a lattice in the vectorspace $U(F)(\mathbb{R})\cong \underline{u}(F)_{\mathbb{R}}$. Choose vectors $e_1,\dots,e_s \in C(F)$ such that $\exp(e_1),\dots,\exp(e_s)$ is a basis of the free group $\Delta=\Gamma\cap U(F)_{\mathbb{R}}$. Identify $\underline{u}(F)_{\mathbb{R}}$ with $\mathbb{R}^s$ via this basis.

Denote by T the torus

$$T = {}_{\Delta}\backslash^{U(F)(\mathbb{C})}$$

$$\cong {}_{\mathbb{Z}^r}\backslash^{\mathbb{C}^r} \xrightarrow[\exp(2\pi i z_j)]{\sim} (\mathbb{C}^x)^r .$$

Then T operates freely on

$${}_{\Delta}\backslash^{D_F} \cong \left({}_{\Delta}\backslash^{\underline{u}}(F)_{\mathbb{C}}\right) \times \underline{v}(F) \times F$$

and this space becomes a principal T-bundle over ${}_{U(F)(\mathbb{C})}\backslash^{D_F}$.

T operates also on $\mathbb{C}^r$, and $T=(\mathbb{C}^x)^r \subseteq \mathbb{C}^r$ is a T-equivariant embedding. We thus may form an embedding

$${}_{\Delta}\backslash^{D_F} \subseteq {}_{\Delta}\backslash^{D_F} \times_T \mathbb{A}^r$$

The second space is a fibre bundle over ${}_{U(F)(\mathbb{C})}\backslash^{D_F}$ with fibre $\mathbb{A}^r$. We need the following fact from [AMRT] and [M3] :

Fact:

a) There exists a compact complex algebraic manifold $X^+ \supseteq X$, such that locally the embedding $X \hookrightarrow X^+$ is isomorphic to one of the embeddings above. (For suitable choices of F and $e_1,\dots,e_r$). X^+ dominates X^*.

b) The vector bundle $t^*_{A/X}$ (defined by the various V's) extends to X^+, such that its natural hermitian metric has logarithmic singularities along $X^+ - X$.

c) The extension to X^+ of the determinant bundle $\Lambda^g t^*_{A/X}$ is the pullback of the ample line-bundle on X^*. (These two bundles are already isomorphic over X, and this isomorphism extends)

The proofs cannot be given here. a) is essentially the content of [AMRT], b) and c) can be found in [M3] (Th.3.1 and Prop. 3.4) We just indicate the essential idea behind b) :

Choose F, $e_1, \ldots, e_s$ as in the construction above. Let $z_j : U(F)_{\mathbb{C}} \to \mathbb{C}$ be the coordinate functions dual to $e_1, \ldots, e_s$. The functions $\zeta_j = \exp(2\pi i z_j)$ form part of a local system of coordinates, and the boundary is defined by $\zeta_1 \cdot \ldots \cdot \zeta_r = 0$. Now the singular behaviour of the metric is determined by a polynomial in the z_j, and the z_j are of logarithmic growth.

Corollary 6.2:

For arbitrary $X = {}_\Gamma\backslash^D$, the metric on one of the ample line-bundles on X^* constructed before (corresponding to the r-th power of the $\Lambda^g V$) has logarithmic singularities along $X^* - X$.

BIBLIOGRAPHY:

[A] M. Artin: Algebraization of formal moduli I in: Global Analysis Princeton Univ. Press, Princeton 1969.

[AMRT] A. Ash.D. Mumford, M. Rapoport, Y.Tai: Smooth compactification of locally symmetric varieteis Math. Sci. Press, Brookline (1975).

[BB] W.L. Baily, A. Borel: Compactification of arithmetic quotients of bounded symmetric domains. Ann. of Math. 84(1966), 442-528.

[DM] P. Deligne, D. Mumford: The irreducibility of the space of curves of a given genus Publ. math. IHES 36(1969), 75-110.

[M1] D. Mumford: Geometric Invariant Theory Springer Verlag, Berlin 1965.

[M2] D. Mumford: Stability of projective varieties Ens. Math. 23(1977), 39-100.

[M3] D. Mumford: Hirzebruch's proportionality theorem in the non-compact case Inven. math. 42(1977), 239-272.

II

H E I G H T S

Gerd Faltings

Contents:

§ 1 The definition

Let K denote a number-field. Classically the height $H(x)$ of a point $x=(x_0:\dots:x_n)\in\mathbb{P}^n(K)$ is defined by

$$H(x)^{[K:\mathbb{Q}]} = \prod_{v\in S} \|(x_0,\dots,x_n)\|_v$$

The product runs over the set S of all places v of K, and $\|(x_0,\dots,x_n)\|_v$ is given by:

$\sup\{|x_j|_v \mid 0\le j\le n\}$, if v is finite,

$(\Sigma |x_j|_v^2)^{\varepsilon_v/2}$, if v is infinite,

where $\varepsilon_v = 1$ or 2 , if v is real/complex .

By the product formula this gives a well-defined function on $\mathbb{P}^n(K)$.

For any extension $K_1\subseteq K_2$ the restriction to $\mathbb{P}^n(K_1)$ of the height-function on $\mathbb{P}^n(K_2)$ is the height-function there. Thus $H(.)$ is defined on $\mathbb{P}^n(\overline{\mathbb{Q}})$. We let $h(x) = \log(H(x))$.

Theorem 1.1:

For $c > 0$ is the number of $x\in\mathbb{P}^n(K)$ with $h(x)\le c$ finite.

Proof:

Let $t=[K:\mathbb{Q}]$, and $\sigma_1,\ldots,\sigma_r:K \to \bar{\mathbb{Q}}$ the different embeddings of K into the algebraic closure of $\mathbb{Q}$. Then $(\sigma_1(x),\ldots,\sigma_r(x))$ defines a $\mathbb{Q}$-rational point in the r-fold symmetric product $S^r(\mathbb{P}^n)$ of $\mathbb{P}^n$. Choose polynomials $F_0 \ldots F_N \in \mathbb{Q}[X_{ij}]$ in the variables X_{ij}, $0 \leq i \leq n$, $1 \leq j \leq r$, multihomogeneous of degree $(d,\ldots,d)$ (that is homogeneous of degree d as a polynomial in $X_{0j},\ldots,X_{nj}$) and symmetric (under the action of $\mathfrak{S}_r$ on the j's), which give an embedding

$$\phi : S^r(\mathbb{P}^n) \hookrightarrow \mathbb{P}^N$$

There exists a constant c_0 with

$$h(\phi(\sigma_1(x),\ldots,\sigma_r(x))) \leq d \cdot r \cdot h(x) + c_0$$

We thus reduce to $K=\mathbb{Q}$.

We may assume that $x_0,\ldots,x_n$ are elements of $\mathbb{Z}$, and that their greatest common divisor is 1 . Then

$$h(x) = \log \sqrt{x_0^2+\ldots+x_n^2} ,$$

and the claim is obvious.

Arakelov has given a new formulation for this definition: Denote by $R \subseteq K$ the ring of integers. A metricized line-bundle on $\mathrm{Spec}(R)$ is a projective R-module P of rank 1 , with hermitian metrics on $P \otimes_R \mathbb{C}$ for any embedding

$K \hookrightarrow \mathbb{C}$. For conjugate complex embeddings the metrics should be equal on P , and thus for $p \in P$ we have norms $\|p\|_v$ for any infinite place v of K .

We define

$$\deg(P, \{\| \ \|_v\}) = \log(\mathrm{order}(P/_{R \cdot p})) - \sum_v \varepsilon_v \log \|p\|_v ,$$

where p is an arbitrary nonzero element of P . (The definition is independant of the choice of p)

To any point $x \in \mathbb{P}^n(K)$ there corresponds a morphism

$$\phi : \mathrm{Spec}(R) \to \mathbb{P}^n_{\mathbb{Z}} .$$

On $\mathbb{P}^n_{\mathbb{Z}}$ we have the line-bundle $\mathcal{O}(1)$, the universal quotient of $\mathcal{O}^{n+1}$:

$$\mathcal{O}^{n+1} \to \mathcal{O}(1) .$$

We thus define a hermitian metric on $\mathcal{O}(1) \otimes_{\mathbb{Z}} \mathbb{C}$ (on $\mathbb{P}^n_{\mathbb{C}}$) by taking the quotient of the standard metric on the constant bundle $\mathcal{O}^{n+1}$.

By pullback $\phi^*\mathcal{O}(1)$ becomes a metricized line-bundle on $\mathrm{Spec}(R)$. An easy calculation shows that

$$h(x) = \frac{1}{[K:\mathbb{Q}]} \cdot \deg(\phi^*\mathcal{O}(1))$$

More general, if X is a separated scheme of finite type over $\mathrm{Spec}(\mathbb{Z})$, $\underline{L}$ a line-bundle on X with a hermitian metric $\| \ \|$ on $\underline{L} \otimes_{\mathbb{Z}} \mathbb{C}$, and $\phi : \mathrm{Spec}(R) \to X$ a morphism defining a point $x \in X(K)$, we let

$$h_{\underline{L}}(x) = \frac{1}{[K:\mathbb{Q}]} \deg(\phi^* L)$$

We then have the following properties:

i) Up to a bounded function, $h_{\underline{L}}(.)$ depends only on the isomorphism class of $\underline{L} \otimes_{\mathbb{Z}} \mathbb{Q}$, as a metricized bundle on $X \otimes_{\mathbb{Z}} \mathbb{Q}$.

ii) If $X \otimes_{\mathbb{Z}} \mathbb{Q}$ is proper over $\mathbb{Q}$, $h_{\underline{L}}(.)$ depends up to a bounded function only on the isomorphism class of $\underline{L} \otimes_{\mathbb{Z}} \mathbb{Q}$

iii) If $X \otimes_{\mathbb{Z}} \mathbb{Q}$ is projective and $\underline{L}$ is ample on $X \otimes_{\mathbb{Z}} \mathbb{Q}$, the number of $x \in X(K)$ with $h_L(x) \leq c$ is finite, for any $c > 0$. (Note that we consider only $x \in X(K)$ which extend to $\phi : \mathrm{Spec}(R) \to X$. If X is proper of $\mathrm{Spec}(\mathbb{Z})$, this is automatic)

Property i) follows from generalities about schemes of finite type of over $\mathrm{Spec}(\mathbb{Z})$. For ii) we have to use that $X(\mathbb{C})$ is compact and so any two hermitian metrics on $L \otimes_{\mathbb{Z}} \mathbb{C}$ are mutually bounded. For iii) we may assume that

$$\underline{L} \otimes_{\mathbb{Z}} \mathbb{Q} \cong \mathcal{O}(1)|X ,$$

for an embedding $X \otimes_{\mathbb{Z}} \mathbb{Q} \hookrightarrow \mathbb{P}^n_{\mathbb{Q}}$ ($h_{\underline{L}}(x)$ is linear in $\underline{L}$)

and the claim holds for $\mathbb{P}^n_{\mathbb{Z}}$ with $\mathcal{O}(1)$

The main advantage of Arakelov's definition is that we may choose the metric on $L \otimes_{\mathbb{Q}} \mathbb{C}$ adapted to our problem. It is a coordinate-free approach.

We need a slight generalization:

Suppose X is proper and normal over $\mathrm{Spec}(\mathbb{Z})$, $Y \subseteq X$ a closed nowhere dense subscheme (defined over $\mathbb{Z}$) and $\underline{L}$ an ample line-bundle on X. We suppose that $\underline{L} \otimes_{\mathbb{Z}} \mathbb{C}$ has a hermitian metric on $(X-Y) \otimes_{\mathbb{Z}} \mathbb{C}$, with logarithmic singularities along Y.

If $x \in X(K)-Y(K)$ we extend as usual to a $\phi : \mathrm{Spec}(R) \to X$, and obtain a metricized line-bundle $\phi^*(\underline{L})$ on $\mathrm{Spec}(R)$.

Let $h_{\underline{L}}(x) = \frac{1}{[K:\mathbb{Q}]} \deg(\phi^*(\underline{L}))$

Theorem 1.2:

The number of points $x \in X(K)-Y(K)$ with $h(x) \leq c$ is finite.

Proof:

We may assume that $X \subseteq \mathbb{P}^n_{\mathbb{Z}}$, $Y \subseteq X$ is the intersection of X with a linear subspace, and $\underline{L}$ the restriction of $\mathcal{O}(1)$ to X. There exist $f_1,\dots,f_r \in \Gamma(\mathbb{P}^n_{\mathbb{Z}}, \mathcal{O}(1))$ with Y as common set of zeros on X.

Let $\| \ \|_1$ denote a hermitian metric on $\underline{L} \otimes_{\mathbb{Z}} \mathbb{C}$ (on all of $X \otimes_{\mathbb{Z}} \mathbb{C}$), and $\tilde{h}_{\underline{L}}(x)$ the corresponding height.

As $x \notin Y(K)$, one of the f_i does not vanish at x, and thus $\phi^*(f_i)$ is a non-zero section of $\phi^*(\underline{L})$. Thus

$$\deg(\phi^*(\underline{L}), \phi^* \| \ \|_1) = \log\left(\mathrm{order}\left(\phi^*(\underline{L}) / R\phi^*(f_i)\right)\right) - \sum_{v \in S_\infty} \varepsilon_v \log \| f_i \|_{1,v} .$$

$$\geq - \sum_{v \in S_\infty} \varepsilon_v \cdot \log \| f_i \|_{1,v}$$

$$= - \sum_\sigma \log \| f_i \|_1 (\sigma(x))$$

The last sum goes over all embeddings $K \hookrightarrow \mathbb{C}$. As $\| f_i \|_1 (z)$ is bounded in $X(\mathbb{C})$, there exists a (independant of i and x) such that

$$- \log \| f_i \|_1 (\sigma(x)) \leq a + \deg(\phi^*(\underline{L}), \phi^* \| \ \|_1) ,$$

for all σ and i with $f_i(x) \neq 0$.

Thus

$$-\log(\max_i \| f_i \|_1 (\sigma(x))) \leq a + [K:\mathbb{Q}] \, \tilde{h}_{\underline{L}}(x) .$$

for all σ and $x \in X(K)-Y(K)$.

As $\| \ \|$ has logarithmic singularities along $Y \otimes_{\mathbb{Z}} \mathbb{C}$, there exist constants $b,c > 0$ with

$$|\log \| g \| (z) - \log \| g \|_1 (z) | \leq$$
$$b + c \cdot \log\{\max[1, -\log(\max_i \| f_i \|_1)(z)]\}$$

Hence we find $d,e > 0$ with

$$|h_L(x) - \tilde{h}_{\underline{L}}(x)| \leq d + e \cdot \log \{\max[1, \tilde{h}_L(x)]\}$$

Thus $\tilde{h}_{\underline{L}}(x)$ remains bounded if $h_L(x)$ does, and this proves the theorem.

§ 3 Néron-Tate heights

We want to demonstrate the use of Arakelov's ideas in a relevant example. Let $S=\mathrm{Spec}(R)$ with $R\subseteq K$ as before, and let A be an abelian variety over K. We also denote by A the Néron-model of A over S, and by A^o its connected component. A and A^o are algebraic groups over S,

$$A(K) = A(R) \supseteq A^o(R) ,$$

and $A^o(R)$ has finite index in $A(R)$.

If $\underline{L}$ is a line-bundle on A (over S), we have the function $h_{\underline{L}}(.)$ on $A(R)$. We want to choose the hermitian metrics at the infinite places in such a way that $h_{\underline{L}}(\ )$ becomes a quadratic function on $A^o(R)$. The quadratic part is by definition the Néron-Tate-height. For any embedding $\sigma:K\hookrightarrow \mathbb{C}$ $A\otimes_R\mathbb{C}$ is a complex torus.

If in general $X/\mathbb{C}$ is a complex torus and $\underline{M}$ a line-bundle on X there exists a hermitian metric on $\underline{M}$ whose curvature is translation-invariant. This metric is unique up to scalars. (The curvature is a (1,1)-form, given locally by $\partial\bar{\partial}\log(||h||^2)$, h a local generator of $\underline{M}$. If we use an translation-invariant Kähler-metric on X, the harmonic forms are translation-invariant, and the metric can be chosen such that its curvature is harmonic). Then the metric satisfies the theorem of the cube:

For any subset $I \subseteq \{1,2,3\}$, there are morphisms

$$p_I : X \times_{\mathbb{C}} X \times_{\mathbb{C}} X \to X$$
$$p_I(x_1, x_2, x_3) = \sum_{j \in I} x_j$$

The theorem of the cube means that

$$\sum_I (-1)^{|I|} p_I^*(\underline{M}) = 0$$

in $\mathrm{Pic}(X \times_{\mathbb{C}} X \times_{\mathbb{C}} X)$, that is

$$\mathcal{O}_{X \times X \times X} \cong \bigotimes_I p_I^*(\underline{M})^{\otimes(-1)^{|I|}}$$

This isomorphism can be normalized in such a way that it is the identity on $\{e\} \times_{\mathbb{C}} X \times_{\mathbb{C}} X$, where $e \in X(\mathbb{C})$ is the neutral element. (The right hand side is canonically trivialized on $\{e\} \times X_{\mathbb{C}} \times X_{\mathbb{C}})$.

If we use the pullbacks by p_I of our hermitian metric on $\underline{M}$, we obtain a hermitian metric on

$$\bigotimes_I p_I^*(\underline{M})^{(-1)^{|I|}}$$

Its curvature is given by

$$\sum_I (-1)^{|I|} p_I^*(\mathrm{curvature}_{\underline{M}})$$

As $\mathrm{curvature}_{\underline{M}}$ is a quadratic function on the tangent space of X, this vanishes. We therefore have obtained a multiple of the standard metric on $\mathcal{O}_{X \times X \times X}$. Using the trivialization on

$\{e\} \times X \times X$ we see that in fact we have an isometry

$$\mathcal{O}_{X \times X \times X} \cong \bigotimes_I p_I^*(\underline{M})^{(-1)^{|I|}}$$

We now go back to arithmetic, and apply this to our bundle $\underline{L}$ on A. For any $\sigma: K \hookrightarrow \mathbb{C}$ we take a hermitain metric on $\underline{L} \otimes_R \mathbb{C}$ with translation invariant curvature, and use these metrics to define $h_{\underline{L}}(.)$

Theorem:

$h_{\underline{L}}(x)$ is a polynomial function on $A^o(R)$, of degree at most two.

proof:

The theorem of the cube gives an isometric isomorphism of bundles on

$A^o \times_S A^o \times_S A^o$:

$$\mathcal{O}_{A^o \times_S A^o \times_S A^o} \cong \bigotimes_I p_I^*(\underline{L})^{(-1)^{|I|}}$$

(At first the right-hand side is trivial on the fibres, hence induced from a bundle on S. Restrict to the zero-section!) Taking degrees this translates into

$$\Sigma(-1)^{|I|} h_{\underline{L}}(p_I(x,y,z)) = 0 ,$$

for $x, y, z \in A^o(R)$.

Thus $h_{\underline{L}}(.)$ is polynomial on $A^o(R)$.

§ 3 Heights on the moduli-space

As before, A_g denotes the coarse moduli-space of principally polarized abelian varieties of dimension g. It is defined over $\mathbb{Q}$, (we do not need it over $\mathbb{Z}$), and there exists a line-bundle $\underline{L}$ on A_g giving the "r'th power of $\omega_{A/A_g} = \Lambda^g t^*_{A/A_g}$", $r > 0$. (As A_g is not a fine moduli-space, there does not exist an universal A).

We assume that $g \geq 2$. If we replace r by a suitable multiple, the sections of $\underline{L} \otimes_{\mathbb{Q}} \mathbb{C}$ give a projective embedding of $A_g \otimes_{\mathbb{Q}} \mathbb{C}$, by the theory of the minimal compactification.
By descent there is an embedding $A_g \hookrightarrow \mathbb{P}^n_{\mathbb{Q}}$ such that $\underline{L} = \mathcal{O}(1) | A_g$. We denote by M the Zariski-closure of A_g in $\mathbb{P}^n_{\mathbb{Z}}$, and by $\underline{L}$ the line-bundle $\mathcal{O}(1) | M$.
Then $M \otimes_{\mathbb{Z}} \mathbb{C} \cong (A_{g,\mathbb{C}})^*$.

The bundle $\underline{L} \otimes_{\mathbb{Z}} \mathbb{C}$ on $A_{g,\mathbb{C}}$ has a natural hermitian metric, defined by square-integration of differentials:
If $A/\mathbb{C}$ is an abelian variety over $\mathbb{C}$, and $\alpha \in \omega_{A/\mathbb{C}} = \Gamma(A, \Omega^g_{A/\mathbb{C}})$,

$$\|\alpha\|^2 = (-1)^{\frac{g(g-1)}{2}} \left(\frac{i}{2}\right)^g \int_{A(\mathbb{C})} \alpha \wedge \bar{\alpha} .$$

Up to a constant factor this metric coincides with the metric on $(\Lambda^g t^*_{A/A_g})^{\otimes r}$ defined in Ch. I, §6. Therefore it has logarithmic singularities at infinity. (Ch. I, Cor 6.2.)
We thus can define a height-function $h_{\underline{L}}$ on $A_g(\bar{\mathbb{Q}})$, such that for number-fields K there are only finitely many

$x \in A_g(K)$ with $h_{\underline{L}}(x) \leq c$, for any c (by Th. 1.2).
The purpose of this chapter is to compute $h_{\underline{L}}(x)$ in case x is the K-rational point defined by a semi-stable principally polarized abelian variety A over K . More precisely, we define a moduli-theoretic height $h(A)$ for such an A , as follows:

Consider the connected component of the Nêron-model of A over R , $A^o \to \mathrm{Spec}(R)$. The bundle $t^*_{A/R}$ has hermitian metrics at the infinite places, and thus $\omega_{A/R} = \Lambda^g t^*_{A/R}$ is a metricized line-bundle over $\mathrm{Spec}(R)$. Let

$$h(A) = \frac{1}{[K:\mathbb{Q}]} \deg(\omega_{A/R}) .$$

Then $h(A)$ is invariant under extensions of K (since $\omega_{A/R}$ is), and we have:

Theorem 3.1:
There exists a constant C, independant of K and A , such that

$$|h_{\underline{L}}(x) - r \cdot h(A)| \leq C .$$

Proof:
There exists a "covering"

$$\phi_i : U_i \to M , \text{ with } U_i \text{ schemes },$$

such that

a) Over $\phi_i^{-1}(A_{g,\mathbb{Q}})$, there exists a universal abelian variety A_i .

b) Over U_i exists a stable curve

$q_i: C_i \to U_i$,

with smooth generic fibre , and morphisms

$\mathrm{Pic}^o(C_i) \underset{\beta_i}{\overset{\alpha_i}{\rightleftarrows}} A_i$, with $\beta_i \circ \alpha_i = d\cdot \mathrm{id}$, $d > 0$.

c) There exist line-bundles

$M_i \subseteq \Lambda^g q_{i,*}(\omega_{C_i/U_i})$,

which are locally direct summands, such that over

$\phi_i^{-1}(A_{g,\mathbb{Q}})$ $\underline{M}_i$ is the image of

$\alpha_i^*: \omega_{A_i/U_i} = \Lambda^g t^*_{A_i/U_i} \to \Lambda^g q_{i,*}(\omega_{C_i/U_i})$

This follows, because we realize the conditions a) b) c) step by step by taking "coverings":
For a) this follows from I, § 2 for b) from I, 3.2/3.3, and for c) we note that $\underline{M}_i$ is already defined over $\phi_i^{-1}(A_{g,\mathbb{Q}})$. This defines a mapping from $\phi_i^{-1}(A_g,\mathbb{Q})$ into a suitable projective bundle, and we take the normalization of the closure of its graph. We further may assume:

d) The isomorphism used to define $\underline{L}$ on A_g :

$\phi_i^*(\underline{L}) \cong \omega_{A_i/U_i}^{\otimes r} \cong \underline{M}_i^{\otimes r}$, over $\phi_i^{-1}(U_i)$,

extends to an isomorphism

$\phi_i^*(\underline{L}) \cong \underline{M}_i^{\otimes r}$

on $U_i \otimes_{\mathbb{Z}} \mathbb{Q}$.

For this claim we may extend from $\mathbb{Q}$ to $\mathbb{C}$. Then both line-bundles carry hermitian metrics with logarithmic singularities along the union of the discriminant locus of C_i and $\phi_i^{-1}(M_{\mathbb{Q}}-A_{g,\mathbb{Q}})$. If we show that the isomorphism between them on $\phi_i^{-1}(A_{g,\mathbb{Q}})$, as well as its inverse, is uniformely bounded, our claim follows. This comes down to the fact that the isomorphism

$$\alpha_i^*:\omega_{A_i/U_i} \otimes_{\mathbb{Z}} \mathbb{C} \xrightarrow{\sim} \underline{M}_i \otimes_{\mathbb{Z}} \mathbb{C}$$

and its inverse are uniformely bounded. Here the metric on

$$\omega_{A/U_i} = \Lambda^g \, t^*_{A_i/U_i}$$

is given by the polarized Hodge-structure corresponding to A_i , while the metric on $\underline{M}_i \subseteq \Lambda^g q_*(\omega_{C_i/U_i})$ comes from the polarization on A_i induced from the polarization on $\mathrm{Pic}^o(C_i)$ by

$$\alpha_i:\mathrm{Pic}^o(C_i) \to A_i$$

As two polarizations on an abelian variety are comparable, we are done.

The rest of the proof is rather easy:
As the U_i are of finite type over $\mathrm{Spec}(\mathbb{Z})$, there exists a number $n > 0$, such that $\phi_i^*(\underline{L})$ and $\underline{M}_i^{\otimes r}$ are "Isomorphic up to a factor n" , that is, if we denote by

$$\gamma_i : \phi_i^*(\underline{L}) \cong \underline{M}_i^{\otimes r}$$

the isomorphism given on $U_i \otimes_{\mathbb{Z}} \mathbb{Q}$, $n\cdot\gamma_i$ and $n\cdot\gamma_i^{-1}$ extend to regular mappings between $\phi_i^*(\underline{L})$ and $\underline{M}_i^{\otimes r}$ on U_i.

Now let A/K be a principally polarized abelian variety over a number-field K, A^o/R its Néron-model, $x \in A_g(K)$ the corresponding moduli-point.

We claim that

$$|h_{\underline{L}}(x) - r\cdot h(A)| \leq \log(n) + r\cdot g\cdot \log(d)$$

(d as in b) above)

For this we may extend K. We then may assume that there exists a Zariski-open cover $\mathrm{Spec}(R) = \bigcup V_i$ and mappings

$$\psi_i : V_i \to U_i ,$$

such that

$$\phi_i \circ \psi_i = \phi | V_i ,$$

where

$$\phi : \mathrm{Spec}(R) \to M$$

is defined by x, and such that the pullback by $\psi_i(K)$ of A_i is isomorphic to A/K.

By pullback, we obtain stable curves $D_i = \psi_i^*(C_i)$ over V_i, and morphisms over $\mathrm{Spec}(K)$

$$\mathrm{Pic}^o(D_i)_{/K} \underset{\beta_i}{\overset{\alpha_i}{\rightleftarrows}} \psi_i^*(A_i)_{/K} \cong A_{/K}$$

$\beta_i \circ \alpha_i = d\cdot \mathrm{id}$.

Furthermore there exists a direct summand

$$\psi_i^*(\underline{M}_i) \subseteq \Lambda^g q_{i,*}(\omega_{D_i/V_i}) ,$$

such that over $\operatorname{Spec}(K_i)$ $\psi_i^*(\underline{M}_i)$ is the image of

$$\alpha_i^* : \omega_{A/K} \to \Lambda^g q_{i,*}(\omega_{D_i/V_i}) .$$

By the theory of minimal models α_i and β_i can be extended to V_i :

$$\operatorname{Pic}^0(D_i) \underset{\beta_i}{\overset{\alpha_i}{\rightleftarrows}} A/V_i ,$$

and $\psi_i^*(\underline{M})$ must be the unique direct summand of $\Lambda^g q_{i,*}(\omega_{D_i/V_i})$ containing the image of α_i^* , so that

$$d^g\, \psi_i^*(\underline{M}_i) \subseteq \alpha_i^*(\omega_{A/V_i}) \subseteq \psi_i^*(\underline{M}_i) \subseteq \Lambda^g q_{i,*}(\omega_{D_i/V_i})$$

Finally, there is a commutative diagram of isomorphisms

$$\begin{array}{ccc} & & (\psi_i \circ \phi_i)^*(\underline{L} \otimes_{\mathbb{Z}} \mathbb{Q}) \\ & & \| \\ (\omega_{A/R} \otimes_R K)^{\otimes r} & \cong & \phi^*(\underline{L} \otimes_{\mathbb{Z}} \mathbb{Q}) \\ & \alpha_i^* \searrow \wr & \wr \downarrow \psi_i^*(\gamma_i) \\ & & (\psi_i^*(\underline{M}_i) \otimes_R K)^{\otimes r} \end{array}$$

The isomorphism at the top comes from $A/K \cong \psi_i^*(A_i)$, and it induces an isometry at the infinite places (after base-change with $\sigma : K \to \mathbb{C}$) . We thus may view $\omega_{A/R}^{\otimes r}$, $\phi^*(\underline{L})$ and

$(\psi_i^*(\underline{M}_i))^{\otimes r}$ as submodules of a fixed one-dimensional vector-space V over K, with hermitian metrics on $V \otimes_K \mathbb{C}$ for any

$$\sigma: K \to \mathbb{C} .$$

$\omega_{A/R}^{\otimes r}$ and $\phi^*(\underline{L})$ are projective of rank 1 over R, and their degrees are $r \cdot h(A)$ and $h_{\underline{L}}(x)$. If $R_i = \Gamma(V_i, \mathcal{O}_{V_i})$, then $\psi_i^*(\underline{M}_i)^{\otimes r}$ is projective of rank 1 over R_i.

We now have:

$$d^{rg} \cdot \psi_i^*(\underline{M}_i)^{\otimes r} \subseteq (\omega_{A/R})^{\otimes r} R_i \subseteq \psi_i^*(\underline{M}_i)^{\otimes r} ,$$

$$n \cdot \psi_i^*(\underline{M}_i)^{\otimes r} \subseteq \phi^*(\underline{L}) \cdot R_i \subseteq n^{-1} \cdot \psi_i^*(\underline{M}_i)^{\otimes r} ,$$

hence

$$n \cdot (\omega_{A/R})^{\otimes r} \cdot R_i \subseteq \phi^*(\underline{L}) \cdot R_i \subseteq d^{-rg} \cdot n^{-1} (\omega_{A/R})^{\otimes r} \cdot R_i$$

As the V_i form a covering of $\operatorname{Spec}(R)$,

$$n \cdot (\omega_{A/R})^{\otimes r} \subseteq \phi^*(\underline{L}) \subseteq d^{-rg} \cdot n^{-1} (\omega_{A/R})^{\otimes r} ,$$

and so indeed

$$|h_{\underline{L}}(x) - r \cdot h(A)| \leq \log(n) + rg \cdot \log(d) .$$

§ 4 Applications

We shall need the following lemma (Hermite-Minkowski)

Lemma 4.1:

Let K be a number-field, S a finite set of places of K. For given $d > 0$ there exist only finitely many extensions $L \supseteq K$ of degree $\leq d$, which are unramified outside S.

Proof:

We first use that a local field of characteristic 0 has only finitely many extensions of degree $\leq d$: This is known for abelian extensions by local classfield-theory, and by induction one reduces to this case because the absolute Galois-group of a local field is solvable.

In the global case this shows that the discriminant of L is bounded. By Minkowski's theorem there exists a constant $C > 0$ and an integral element $x \in L$ with $|x|_{v_1} \leq C$, $|x|_{v_2} < 1$, $|x|_{v_r} < 1$, where the v_i denote the infinite places of L. The coefficients of the minimal polynomial of x are bounded, so that there exist only finitely many possibilities for this polynomial and for $K(x)$. Now $[L:K(x)] \leq 2$, and we may assume that L is a quadratic extension of $K(x)$. By classfield theory there are only finitely many such extensions which are unramified outside S.

We use this lemma in the following form:

Lemma 4.2

Let K be a number-field, S a finite set of places of K.

There exists a finite extension $K' \supseteq K$, such that for any abelian variety A over K of dimension g, with good reduction outside S, the abelian variety $A \otimes_K K'$ is semistable, and has a level-12-structure. (All its 12-division points are rational over K')

<u>proof:</u>

For any such A, the field $K(A[12])$ obtained by adjoining the 12-division points is unramified over K outside S and places of characteristics 2 or 3, and of degree $\leq 12^{4g}$ over K. Hence there exists a K' containing all such $K(A[12])$. As any abelian variety with a level-12-structure is semistable, we are done.

<u>Remark:</u>

We have used the following fact: Any automorphism of finite order of $\mathbb{Z}_l^r$ ($\mathbb{Z}_l$=l-adic integers) which is the identity mod 4 (for l=2) or mod l (for $l \geq 3$) is the identity.

Now follows the main result of the first two exposées:

<u>Theorem 4.3:</u>

Let K be a number-field. Fix an integer $g \geq 2$ and a $c>0$. There exist up to isomorphism only finitely many principally polarized semistable abelian varieties A over K, such that $h(A) \leq c$.

<u>Proof:</u>

Let $x \in A_g(K)$ be the moduli-point for such an A. We have seen that $|h_{\underline{L}}(x) - rh(A)|$ is bounded, so that we obtain only finitely many different x. If two A's give the same x, they become isomorphic over the algebraic closure $\bar{K}$ of K, hence over a finite extension of K. They then have bad reduction at the same places of K.

By the previous lemma there exists a finite Galois-extension $K' \supseteq K$ such that all the A's have rational 12-division-points over K'. Any isomorphism between them over a finite extension of K' then is already defined over K' itself, since the isomorphism is already determined by its effect on 12-torsion-points, and hence equal to its Galois-conjugates. Thus all A's inducing the same $x \in A_g(K)$ become isomorphic over K'. They are then parametrized by a subset of the finite set

$$H^1\big(\mathrm{Gal}(K'/K), \mathrm{Aut}(A/K', \text{polarization})\big) .$$

This proves our claim.

Remark:

Theorem 4.3 holds also for isomorphism classes of abelian varieties (forgetting polarizations). See Ch. IV, Lemma 3.8.

III

SOME FACTS FROM THE THEORY OF GROUP SCHEMES

Fritz Grunewald

Contents:

§0 Introduction

This paper discusses some results which are used in the contributions of Schappacher and Wüstholz to this volume. I have tried to explain the application of the theories of finite group schemes and p-divisible groups to the problems arising in Faltings work.

Where it seemed necessary and where it was possible for me, I have given detailed proofs. I have also included many examples.

Chapters one and two introduce to the theory of group schemes in particular finite group schemes. Most important are here the exactness properties of the functor $s^*\Omega^1$.

Chapter three discusses p-divisible groups, a concept introduced by Tate.

In chapters four and five we study the action of the absolute galois group on the points of a finite commutative group scheme and on the Tate-module of a p-divisible group.

I thank G. Faltings who has helped me a lot with writing this paper.

§1 Generalities on group schemes

In this paragraph we describe certain elementary facts from the theory of group schemes. We shall use the language of schemes as set up for example in [H].

Let $\underline{S}$ be a fixed scheme, then the category of schemes over $\underline{S}$ has a categorial product which comes from the usual fibre product of schemes.

$$\begin{array}{ccc} \underline{X}\times_{\underline{S}}\underline{Y} & \longrightarrow & \underline{X} \\ \downarrow & & \downarrow \\ \underline{Y} & \longrightarrow & \underline{S} \end{array}$$

So we have the notion of a group object in the category of schemes over $\underline{S}$. A group scheme over $\underline{S}$ is then a map of schemes

$$\begin{array}{c} \underline{G} \\ \downarrow \\ \underline{S} \end{array}$$

together with maps of schemes over $\underline{S}$:

$$\mu: \quad \begin{array}{ccc} \underline{G}\times_{\underline{S}}\underline{G} & \longrightarrow & G \\ \downarrow & & \downarrow \\ \underline{S} & \longrightarrow & \underline{S} \end{array}$$

$$s: \quad \begin{array}{ccc} \underline{S} & \longrightarrow & \underline{G} \\ \downarrow & \text{id} & \downarrow \\ \underline{S} & \longrightarrow & \underline{S} \end{array}$$

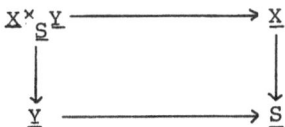

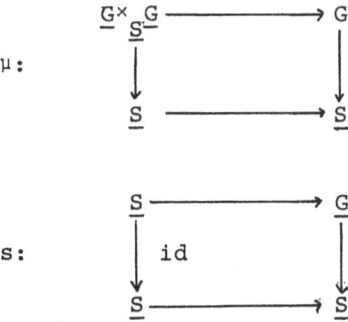

i: 

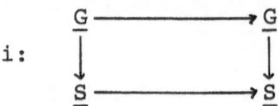

such that the following diagrams are commutative:

1) 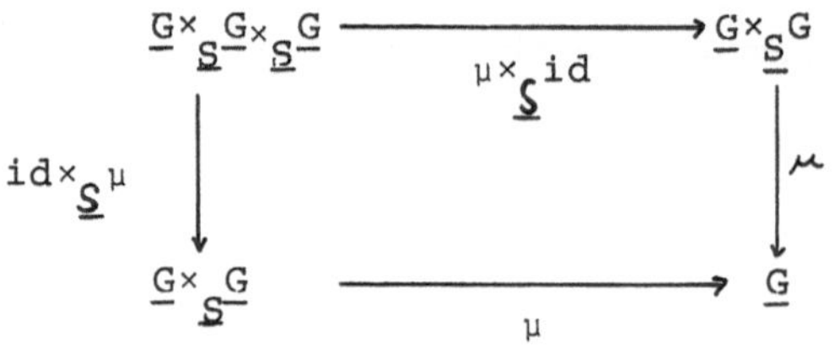

2) 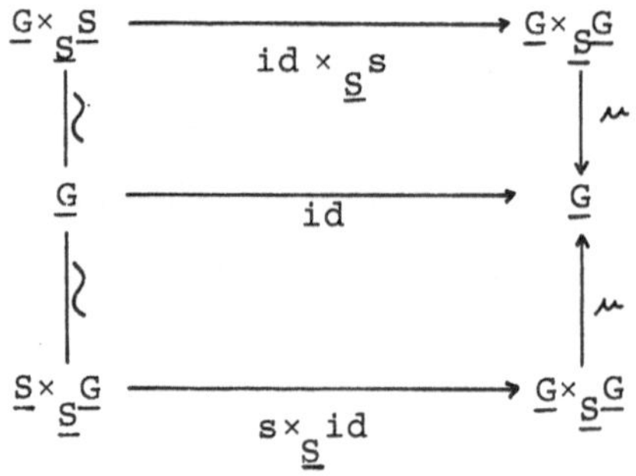

3) 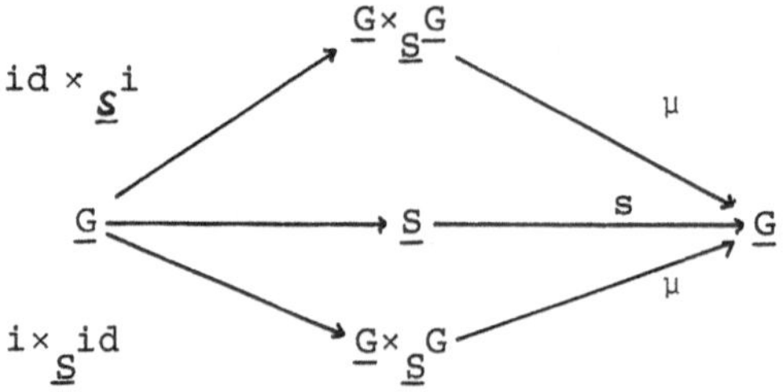

A group scheme $G \to S$ is called commutative if the following diagram

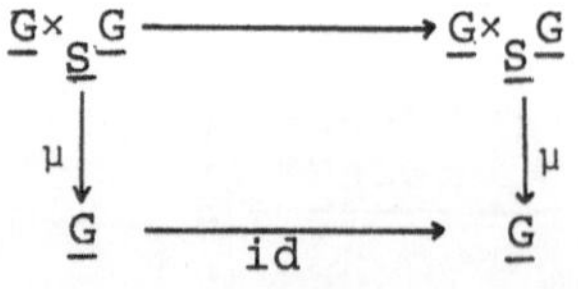

is commutative. Here σ is the map which interchanges the components of the product. Let $\underline{G} \to \underline{S}$ a group scheme and $\underline{T} \to \underline{S}$ a scheme over $\underline{S}$, then the structural maps of $\underline{G} \to \underline{S}$ induce on

$$\underline{G}(\underline{T}) = \mathrm{Hom}_{\underline{S}}\left(\begin{array}{c}\underline{T}\\ \downarrow\\ \underline{S}\end{array}, \begin{array}{c}\underline{G}\\ \downarrow\\ \underline{S}\end{array}\right)$$

a group structure. $\underline{G}(\underline{T})$ is called the group of $\underline{T}$-valued points of $\underline{G}$. Let $\underline{G} \to \underline{S}$ and $\underline{H} \to \underline{S}$ be group schemes over $\underline{S}$. A map of schemes over $\underline{S}$

$$\varphi: \begin{array}{ccc}\underline{G} & \longrightarrow & \underline{H}\\ \downarrow & & \downarrow\\ \underline{S} & \longrightarrow & \underline{S}\end{array}$$

is called a homomorphism if the following diagram is commutatative

$$\begin{array}{ccc}\underline{G}\times_{\underline{S}}\underline{G} & \xrightarrow{\varphi\times_{\underline{S}}\varphi} & \underline{H}\times_{\underline{S}}\underline{H}\\ \mu\downarrow & & \downarrow\mu\\ \underline{S} & \xrightarrow[\mathrm{id}]{} & \underline{S}\end{array} .$$

If $\varphi: (\underline{G} \to \underline{S}) \to (\underline{H} \to \underline{S})$ is a homomorphism of group schemes, then the kernel of φ is the fibre product of the following diagram

$$\begin{array}{ccc}\underline{K} & \longrightarrow & \underline{S}\\ \downarrow & & \downarrow s\\ \underline{G} & \xrightarrow[\varphi]{} & \underline{H}\end{array} .$$

The structural maps of $\underline{G}$ induce on $\underline{K} \to \underline{S}$ a group scheme structure.

We shall mostly consider the case where the base scheme $\underline{S}$ is affine, that is $\underline{S}$ is of the form spec(R) for some commutative ring R. A scheme $\underline{X} \to \text{spec}(R)$ is called a scheme defined over R. Consider the case where $\underline{X}$ is also affine, $\underline{X} = \text{spec}(A)$. The map $\underline{X} \to \text{spec}(R)$ comes from a ring homomorphism. $R \to A$. Let now A be an R-algebra. The structural maps of a group scheme on $\text{spec}(A) \to \text{spec}(R)$ come from R-algebra homomorphisms:

$$\mu: \quad A \to A \otimes_R A$$

$$s: \quad A \to R$$

$$i: \quad A \to A.$$

The maps μ, s, i make certain obvious diagrams commutative. Conversely given an R-algebra A and maps μ, s, i making the appropriate diagrams commutative one gets on $\text{spec}(A) \to \text{spec}(R)$ the structure of a group scheme. An R-algebra A together with R-Algebra homomorphisms μ, s, i satisfying the appropriate conditions is called a bigebra in [Bo].

Examples:

We shall now give some examples of group schemes over a ring R. They will all be affine. We shall describe them by giving the R-algebra homomorphisms corresponding to μ, s, i.

Example 1: The additive group $\underline{G}_a$.

$A = R[t]$

$\mu: t \to 1 \otimes t + t \otimes 1$

$s: t \to 0$

$i: t \to -t.$

If B is an R-algebra then there is a group isomorphism

$$\underline{G}_a(\text{spec}(B)) \cong B^+.$$

B^+ is the additive group of B. If $\underline{G} \to \text{spec}(R)$ is a group scheme and B is an R-algebra we write

$$\underline{G}(\text{spec}(B)) =: \underline{G}(B)$$

for the group of B-valued points.

Example 2: The multiplicative group $\underline{G}_m$.

$A = R[t,t^{-1}]$

$\mu: t \to t \otimes t$

$s: t \to 1$

$i: t \to t^{-1}$.

If B is an R-algebra then there is a group isomorphism

$$\underline{G}_m(B) \cong B^*.$$

B^* is the group of units in B.

Example 3: The group of n-th roots of unity μ_n. For $n \in \mathbb{N}$ put

$A = R[t,t^{-1}]_{/<t^n-1>}$

$\mu: t \to t \otimes t$

$s: t \to 1$

$i: t \to t^{-1}$.

Example 4: The constant group scheme $\mathcal{K}(\Delta)$. For a group Δ put $A = R^\Delta$ where R^Δ is the ring of R valued functions on Δ.

μ: $f \to \mu f$ with $\mu f(g,h) = f(g \cdot h)$

s: $f \to f(1)$

i: $f \to if$ with $if(g) = f(g^{-1})$.

The schemes in example 4 are all étale over R

Example 5: $\underline{G}_{a,b}$. For $a,b \in R$ with $a \cdot b = 2$ define

$$A = R[t]/_{<t^2-at>}$$

$$\mu: t \to 1 \otimes t + t \otimes 1 - bt \otimes t$$

$$s: t \to 0$$

$$i: t \to -t.$$

The R-algebra A is a two dimensional free R-modul.

Example 6: Many examples arise from (affine) algebraic groups over fields. Let R be a ring with quotient field K. A group scheme $\underline{G}$ over spec(R) is called an abelian scheme over R if $\underline{G}$ is proper and smooth over spec(R) and if all fibres are connected.

Exact sequences:

Definition: Let $\underline{G}_1$ $\underline{G}_2$, $\underline{G}_3$ be group schemes over $\underline{S}$. A sequence of homomorphisms

$$0 \to \underline{G}_1 \overset{\varphi}{\to} \underline{G}_2 \overset{\psi}{\to} \underline{G}_3 \to 0$$

is called exact if

1) φ is a closed immersion identifying $\underline{G}_1$ with the kernel of ψ.

2) ψ is faithfully flat.

Assume that ψ is of finite type, then condition 2 implies that ψ is a strict epimorphism. That means that the sequence

$$\underline{G}_2 \times_{\underline{G}_3} \underline{G}_2 \rightrightarrows \underline{G}_2 \xrightarrow{\psi} \underline{G}_3$$

is exact in the category of schemes. See [M] Theorem 2.17. Assume that $\underline{S}$ and $\underline{G}_1, \underline{G}_2, \underline{G}_3$ are affine, say $\underline{S} = \mathrm{spec}(R)$, $\underline{G}_i = \mathrm{spec}(A_i)$, $i = 1,2,3$. Then the above sequence comes from a sequence of R-algebra homomorphisms

$$A_1 \xleftarrow{\tilde{\varphi}} A_2 \xleftarrow{\tilde{\psi}} A_3 .$$

Condition 2 means that A_2 is under $\tilde{\psi}$ a faithfully flat A_3-module. Condition 1 means that $\tilde{\varphi}$ is surjective and that there is an R-algebra isomorphism

$$\Theta : A_2 \otimes_{A_3} R \longrightarrow A_1$$

making the following diagram commutative

$$\begin{array}{ccccc} & & A_2 \otimes_{A_3} R & \longleftarrow & R \\ & \swarrow^{\Theta} & \uparrow & & \uparrow s \\ A_1 & \xleftarrow{\tilde{\varphi}} & A_2 & \xleftarrow{\tilde{\psi}} & A_3 \end{array} .$$

The tensor product is formed viewing R as an A_3-algebra under the zero-section s.

<u>Modules of differentials</u>:

We consider here the case where $\underline{S} = \mathrm{spec}(R)$ and where $\underline{G} = \mathrm{spec}(A)$ is an affine group scheme over $\underline{S}$. We write

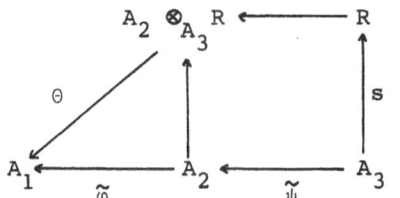

$$\Omega^1_{\underline{G}/\underline{S}} = \Omega^1_{\underline{G}/R} = \Omega^1_{A/R} ,$$

where $\Omega^1_{A/R}$ is the usual A-module of Kähler-differentials of the R-algebra A. See [H],[Gr] for the definitions. The universal derivation is

$$d\colon A \to \Omega^1_{A/R}.$$

We shall also be interested in the following R-module

$$s^*\Omega^1_{A/R} = \Omega^1_{A/R} \otimes_A R.$$

Here the tensorproduct is formed over the zero-section $s\colon A \to R$. For later computation we need the following result.

Proposition 1.1: Let $0 \to \underline{G}_1 \xrightarrow{\varphi} \underline{G}_2 \xrightarrow{\psi} \underline{G}_3$ be an exact sequence of affine group schemes over a ring R. Let $\underline{G}_i = \operatorname{spec}(A_i)$ for $i = 1,2,3$. Then the sequence

$$0 \leftarrow \Omega^1_{\underline{G}_1/R} \leftarrow \Omega^1_{\underline{G}_2/R} \otimes_{A_2} A_1 \leftarrow \Omega^1_{\underline{G}_3/R} \otimes_{A_3} A_1$$

of A_1-modules is exact.

Remarks: 1) From the result in proposition 1.1 it follows that the following sequence of R-modules is also exact:

$$0 \longrightarrow s^*\Omega^1_{\underline{G}_1/R} \longrightarrow s^*\Omega^1_{\underline{G}_2/R} \longrightarrow s^*\Omega^1_{\underline{G}_3/R}.$$

2) The above sequence of group schemes is exact if condition (1) in the definition for exactness of short exact sequences is satisfied.

Proof: Consider the underlying sequence of R-algebras

$$A_1 \xleftarrow{\varphi} A_2 \xleftarrow{\psi} A_3.$$

Since φ is a closed immersion φ has to be surjective. But then the map induced by φ

$$\Omega^1_{A_1/R} \longleftarrow \Omega^1_{A_2/R} \otimes_{A_2} A_1$$

is surjective. Next, we take the sequence $A_2 \xleftarrow{\psi} A_3 \longleftarrow R$ of rings and get, using the second exact sequence, an exact sequence of A_2-modules

$$0 \longleftarrow \Omega^1_{A_2/A_3} \longleftarrow \Omega^1_{A_2/R} \longleftarrow \Omega^1_{A_3/R} \otimes_{A_3} A_2 \, .$$

We tensor this sequence with A_1 and obtain

$$0 \longleftarrow \Omega^1_{A_2/A_3} \otimes_{A_2} A_1 \longleftarrow \Omega^1_{A_2/R} \otimes_{A_2} A_1 \longleftarrow \Omega^1_{A_3/R} \otimes_{A_3} A_1 \, .$$

Using the commutative diagram of R-algebras

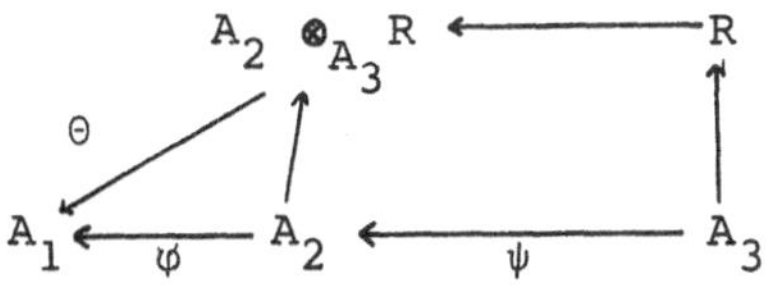

we get a commutative diagram

$$0 \longleftarrow \Omega^1_{A_2/A_3} \otimes_{A_2} A_1 \longleftarrow \Omega^1_{A_2/R} \otimes_{A_2} A_1 \longleftarrow \Omega^1_{A_3/R} \otimes_{A_3} A_1$$

$$\uparrow \varepsilon \qquad\qquad \downarrow \varphi$$

$$\Omega^1_{A_2} \otimes_{A_3} R/R \xleftarrow{\theta^{-1}} \Omega^1_{A_1/R}$$

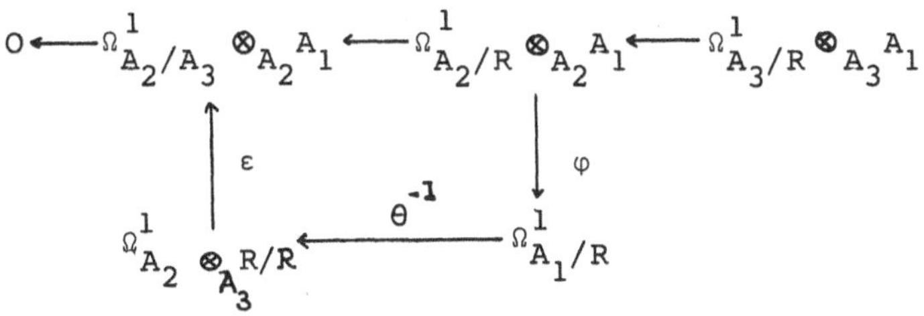

ε is the usual base change isomorphism. Since $\varepsilon \circ \theta^{-1}$ is an isomorphism, proposition 1.1 is proved. $\square$

<u>Remark</u>: If $\underline{Y} \to \underline{X}$ is a map of schemes, then we write

$$\Omega^1_{\underline{Y}/\underline{X}}$$

for the relative module of differentials. Using the same proof, one sees that the assumption in proposition 1.1 that the $\underline{G}_i$ should be affine, is not necessary.

§2 Finite group schemes

Here we shall consider finite group schemes. They arise for example as kernels of isogenies of abelian varieties.

Let $\underline{X}$ be a scheme. For $U \leq \underline{X}^t$, an open set in the underlying topological space of $\underline{X}$, let $\mathcal{O}_{\underline{X}}(U)$ be the associated ring. A morphism $\varphi: \underline{X} \to \underline{S}$ of schemes is called affine if the inverse image under φ of any affine open subset is affine in $\underline{S}$.

Definition: A morphism of schemes $\varphi: \underline{X} \to \underline{S}$ is called finite if it is affine and if for every open affine set $U \subseteq \underline{S}^t$ the $\mathcal{O}_{\underline{S}}(U)$ - algebra $\mathcal{O}_{\underline{X}}(\varphi^{-1}(U))$ is a finitely generated $\mathcal{O}_{\underline{S}}(U)$-module. The morphism φ is called of finite order if the $\mathcal{O}_{\underline{S}}(U)$-modules $\mathcal{O}_{\underline{X}}(\varphi^{-1}(U))$ are locally free of constant rank. If n is this rank then n is called the order of φ. One also says then that $\underline{X}$ is of finite order over $\underline{S}$.

If $\underline{S}$ is a locally noetherian connected scheme then a scheme $\underline{X} \to \underline{S}$ over $\underline{S}$ is of finite order if and only if it is finite and flat over $\underline{S}$. More specifically, consider the case where $\underline{S} = \mathrm{spec}(R)$ for a noetherian local ring R. Let φ: $\underline{X} \to \mathrm{spec}(R)$ be a scheme over $\mathrm{Spec}(R)$. Then $\underline{X}$ is of finite order over $\mathrm{Spec}(R)$ if and only if $\underline{X} = \mathrm{spec}(A)$ for some R-algebra A which is a finitely generated free R-module.

A group scheme $\varphi: \underline{G} \to \underline{S}$ is called finite or of finite order if the map φ is finite or of finite order. The examples 3,5 from §1 are group schemes of finite order over R. The constant group scheme $\mathcal{K}(\Delta)$, this is example 4, is of finite order if the group Δ is finite.

A Theorem of Oort and Tate :

We shall report here on the construction of certain group schemes of prime order due to Oort and Tate [O].

Let p be a prime number. Let ζ be the primitive $(p-1)$-th root of unity in the ring of p-adic integers $\mathbb{Z}_p$ which satisfies $\zeta^m \equiv m \bmod p$ for $m = 1,\dots,p-1$. Put

$$\Lambda_p = \mathbb{Z}\,[\zeta, \frac{1}{p(p-1)}] \cap \mathbb{Z}_p \subseteq \mathbb{Q}_p .$$

We shall construct now certain elements $w_1,\dots,w_{p-1} \in \Lambda_p$. To do this let

$$B = \Lambda_p[z]/_{<z^p-1>}$$

and define in B:

$$w_i = \frac{\left(\sum_{m=1}^{p-1} \zeta^{-m}(1-z^m)\right)^i}{\left(\sum_{m=1}^{p-1} \zeta^{-im}(1-z^m)\right)} .$$

The claim here is that the w_i are units in Λ_p. Examples are easily computed:

$p = 2$: $w_1 = 1$

$p = 3$: $w_1 = 1$, $w_2 = -1$

$p = 5$: $w_1 = 1$, $w_2 = -\zeta(2+\zeta)$, $w_3 = (2+\zeta)^2$,
$w_4 = -5(2+\zeta)^2$.

Take now any Λ_p - algebra R with structural map $\varphi: \Lambda_p \to R$. For any pair $a,b \in R$ with $a \cdot b = p$ define

$$G^p_{a,b} \quad :$$

$$A = R[t]/_{<t^p - at>}$$

$$\mu: t \to t \otimes 1 + 1 \otimes t + \frac{b}{1-p} \sum_{i=1}^{p-1} \frac{1}{\varphi(w_i w_{p-i})} t^i \otimes t^{p-i}$$

$$s: t \to 0$$

$$i: t \to -t \ .$$

Then (A,μ,s,i) is a commutative group scheme of order p over R. This can be checked by computation. Let R now be a complete noetherian local ring of residue characteristic p. R is in a natural way a $\mathbb{Z}_p$-, hence Λ_p-algebra. In this case, we have a group scheme $G^p_{a,b}$ for any pair of elements $a,b \in R$ with $a \cdot b = p$. The following is proved in [O].

Theorem 2.1: (Oort, Tate)

Let R be a complete noetherian local ring of residue characteristic $p > 0$. For any group scheme $\underline{G}$ over R which is finite of order p there are $a,b \in R$ with $a \cdot b = p$ such that $\underline{G}$ and $\underline{G}^p_{a,b}$ are isomorphic as group schemes over R.

Let a,b,c,d be elements of R with $a \cdot b = p$ and $c \cdot d = p$. Then $\underline{G}^p_{a,b}$ and $\underline{G}^p_{c,d}$ are isomorphic if and only if there is a unit $u \in R^*$ with

$$c = u^{p-1}a, \quad d = u^{1-p}b.$$

Note that this theorem implies for certain rings that any group scheme of prime order is commutative. This is proved without restriction on the base scheme in [O]. The $\underline{G}^2_{a,b}$ have already shown up in example 5 of §1.

Duality:

Let R be a ring and $\underline{G} = \mathrm{spec}(A) \to \mathrm{spec}(R)$ a commutative affine finite group scheme over R. $\underline{G}'$ stands for the Cartier dual of $\underline{G}$. It is defined as follows.

$$\underline{G}' = \mathrm{spec}(A')$$

where $A' = \mathrm{Hom}_R(A,R)$. In the Hom only R-module homomorphisms are considered. The structural maps μ, s, i induce maps μ', s', i' which make $\underline{G}'$ into a commutative group scheme over R. If $\underline{G}$ was of finite order then $\underline{G}'$ is also of finite order and the orders coincide. We have

Proposition 2.2: Let p be a prime and R an Λ_p-algebra, and let $a,b \in R$ with $a \cdot b = p$. Then

$$(\underline{G}^p_{a,b})' \cong \underline{G}^p_{b,a} .$$

This can be seen by a straightforward computation, see also [O].

Modules of differentials:

We shall compute now the modules of differentials for the group schemes $\underline{G}^p_{a,b}$. We deduce then some general results on the modules of differentials for group schemes of prime order.

Proposition 2.3: Let p be a prime and R an Λ_p-algebra. For $a,b \in R$ with $a \cdot b = p$ let $\underline{G} = \underline{G}^p_{a,b}$ be the group scheme over R defined above, then

1) $\Omega^1_{\underline{G}/R} = {}^{R[t]}/_{<t^p - at,\, pt^{p-1} - a>}$

2) $s^*\Omega^1_{\underline{G}/R} = {}^{R}/_{a\cdot R}$.

Proof: We have

$$\underline{G} = \underline{G}^p_{a,b} = \mathrm{spec}(A) ,$$

where

$$A = {}^{R[t]}/_{<t^p - at>} .$$

The module of differentials of a polynomial ring is a free one dimensional module:

$$\Omega^1_{R[x]/R} = R[x]\cdot dx$$

with derivation:

$$d: \quad R[x] \to \Omega^1_{R[x]/R}$$

$$d: \quad P(x) \to P'(x)dx$$

From the second exact sequence (1) follows. (2) is proved using the explicit description of the zero section of $\underline{G}^p_{a,b}$. □

Proposition 2.4:

Let R be a complete noetherian local ring of residue characteristic $p > 0$ and without zero divisors. Let $\underline{G}$ be a group scheme of order p over R. Then:

$$\#(s^*\Omega^1_{\underline{G}/R})\cdot \#(s^*\Omega^1_{\underline{G}'/R}) = \#({}^{R}/_{pR})$$

Proof: By theorem 2.1 we find $a,b \in R$ with $a \cdot b = p$ such that

$$\underline{G} \cong \underline{G}^p_{a,b}\ ,\ \underline{G}' \cong \underline{G}^p_{b,a}$$

as group schemes over R. We know by proposition 2.3 that

$$\#(s^*\Omega^1_{\underline{G}/R}) = \#({}^R/_{a\,R}) \quad \text{and}$$

$$\#(s^*\Omega^1_{\underline{G}'/R}) = \#({}^R/_{bR}).$$

We have the exact sequence of R-modules

$$0 \to R/_{aR} \to {}^R/_{a\,bR} \to {}^R/_{bR} \to 0 .$$

From this the result follows. □

A group scheme $\underline{G}$ over a ring R of finite order is of multiplicative type if and only if its dual $\underline{G}'$ is étale over R. For example the schemes μ_n are of multiplicative type. We have

$$(\mu_n)' = \mathfrak{K}({}^{\mathbb{Z}}/_{n\mathbb{Z}}) .$$

Proposition 2.5: Let R be a complete noetherian ring of residue characteristic $p > 0$. Let $\underline{G}$ be a group scheme of order p over R which is of multiplicative type. Then

$$\#(s^*\Omega^1_{\underline{G}/R}) = \#({}^R/_{pR}).$$

Proof: Since $\underline{G}'$ is étale over R we have

$$(s^*\Omega^1_{\underline{G}'/R}) = 0.$$

We then apply proposition 2.4. □

<u>Remark:</u> Proposition 2.4 will be generalised greatly in theorem 2.10.

<u>Étale groups</u>:

Let R be a complete noetherian local ring with quotient field K and residue field k. $\hat{k}$ is the separable algebraic closure of k and $\mathfrak{g}_0$ is its galoisgroup. $R_{\acute{e}t}$ is the maximal local étale extension of R. $\mathfrak{g}_0$ acts naturally on $R_{\acute{e}t}$.

Let M be a finite $\mathfrak{g}_0$-module. Put

$$A = \mathrm{Map}_{\mathfrak{g}_0}(M, R_{\acute{e}t})$$

for the R-algebra of $\mathfrak{g}_0$-invariant $R_{\acute{e}t}$ - valued functions. Define the structural maps for A just as for constant groups. This turns A into an étale bigebra over R. We call this group scheme of finite order $\underline{M}$.

<u>Theorem 2.6</u>: Let R be a complete local ring. The map $M \mapsto \underline{M}$ is an equivalence between the category of finite $\mathfrak{g}_0$-modules and the category of étale group schemes of finite order over R. The inverse map to $M \mapsto \underline{M}$ is given by

$$\underline{G} \mapsto (\underline{G} \otimes_R \hat{k})(\hat{k}).$$

Here

$$\underline{G} \otimes_R \hat{k} = \underline{G} \times_{\mathrm{spec}(R)} \mathrm{spec}(\hat{k}).$$

The fibre product is taken over the map $R \to k \to \hat{k}$. For all of this see [G,D],section II.

We also mention for later use that any group scheme of finite order $\underline{G}$ can be embedded in an exact sequence

$$0 \to \underline{G}^{0} \to \underline{G} \to \underline{G}^{ét} \to 0.$$

where $\underline{G}^{0}$ is a connected subgroup and $\underline{G}^{ét}$ is étale. $\underline{G}^{0}$ and $\underline{G}^{ét}$ are unique up to isomorphism. See [G], [Ra].

<u>Finite subgroups of abelian schemes</u>:

If $\underline{A}$ is an abelian scheme over a ring R then one knows that

$$s^{*}\Omega^{1}_{\underline{A}/R} \cong R^{g}$$

for some $g \in \mathbb{N}$. g is the dimension of $\underline{A}$. See [M]. Assume that $\underline{G}$ is a flat subgroup of finite order in A

$$0 \to \underline{G} \to \underline{A} .$$

Then the exact sequence from proposition 1.1 gives some restriction on $s^{*}\Omega^{1}_{\underline{G}/R}$.

<u>Proposition 2.7</u>: Let K be an algebraic number field of degree m over $\mathbb{Q}$. Let $\underline{A}$ be an abelian scheme of dimension g over the ring of integers $\mathcal{O}$ in K. Let $\underline{G}$ be a flat subgroup of $\underline{A}$ annihilated by a prime number p. Then

$$\#(s^{*}\Omega^{1}_{\underline{G}/R}) = p^{d}$$

with $d \leq m \cdot g$.

<u>Proof</u>: That $\underline{G}$ is annihilated by p means that

multiplication by p factors through the zero section of $\underline{G}$. See the beginning of §3. The map: multiplication by p induces on $s^*\Omega^1_{\underline{G}/R}$ also the multiplication by p. So $s^*\Omega^1_{\underline{G}/R}$ is an abelian group of exponent p. It is also a quotient of $\mathcal{O}^g = \mathbb{Z}^{mg}$. □

We shall also need the following:

Theorem 2.8 (Raynaud): Let R be a local noetherian ring and let $\underline{G}$ be a finite flat group scheme over R. Then there is a projective abelian scheme $\underline{A}$ and a closed immersion

$$0 \to G \to \underline{A}.$$

For this see [Be] p. 110 and [Oo] chapter II for a somewhat weaker version. If $\underline{G}$ is a finite flat subgroup of an abelian scheme $\underline{A}$ then there is an exact sequence

$$\begin{array}{c} 0 \to \underline{G} \to \underline{A} \to \underline{B} \to 0 \\ \qquad\quad\;\; \wr\| \\ \qquad\quad\;\; \underline{A}/\underline{G} \end{array} .$$

This is proved in [M-F].

The exactness of $s^*\Omega^1$:

Here we shall improve on proposition 1.1.

Theorem 2.9: Let R be a discrete valuation ring with quotient field K of characteristic 0. Let

$$0 \to \underline{G}_1 \to \underline{G}_2 \to \underline{G}_3 \to 0$$

be an exact sequence of group schemes of finite order over R. Then the sequence of R-modules

$$0 \to s^*\Omega^1_{\underline{G}_3/R} \to s^*\Omega^1_{\underline{G}_2/R} \to s^*\Omega^1_{\underline{G}_1/R} \to 0$$

is exact.

Proof: The problem here is the injectivity on the left. Let the sequence

$$0 \to G \to \underline{A} \xrightarrow{\varphi} \underline{B} \to 0$$

be exact, where $\underline{G}$ is of finite order and $\underline{A}$, $\underline{B}$ are abelian schemes. The sequence

$$0 \to s^*\Omega^1_{\underline{B}/R} \xrightarrow{\tilde{\varphi}} s^*\Omega^1_{\underline{A}/R} \to s^*\Omega^1_{\underline{G}/R} \to 0$$

is then also exact. This is clear apart from the injectivity on the left. φ is an isogeny and

$$\begin{array}{ccc} \tilde{\varphi}: s^*\Omega^1_{\underline{B}/R} & \longrightarrow & s^*\Omega^1_{\underline{A}/R} \\ \| \wr & & \| \wr \\ R^g & & R^g \end{array}$$

has the degree of φ as determinant. Since K is of characteristic 0 the map $\tilde{\varphi}$ is injective. By theorem 2.8 we embed $\underline{G}_2$ into an abelian scheme $\underline{A}$ and define

$$\underline{B} = \underline{A}/\underline{G}_1 \qquad \underline{C} = \underline{A}/\underline{G}_2$$

Then we have the exact sequences

$$\begin{array}{ccccccccc} & & 0 & & & & & & \\ & & \downarrow & & & & & & \\ 0 & \to & \underline{G}_1 & \to & \underline{A} & \to & \underline{B} & \to & 0 \\ & & \downarrow & & & & & & \\ 0 & \to & \underline{G}_2 & \to & \underline{A} & \to & \underline{C} & \to & 0 \\ & & \downarrow & & & & & & \\ 0 & \to & \underline{G}_3 & \to & \underline{B} & \to & \underline{C} & \to & 0 \\ & & \downarrow & & & & & & \\ & & 0 & & & & & & \end{array}$$

From these we obtain a commutative diagram

$$\begin{array}{ccccccccc}
 & & & & 0 & & & & \\
 & & & & \downarrow & & & & \\
0 & \to & s^*\Omega^1_{\underline{C}/R} & \to & s^*\Omega^1_{\underline{B}/R} & \to & s^*\Omega^1_{\underline{G}_3/R} & \to & 0 \\
 & & \| & & \downarrow & & \downarrow \alpha & & \\
0 & \to & s^*\Omega^1_{\underline{C}/R} & \to & s^*\Omega^1_{\underline{A}/R} & \to & s^*\Omega^1_{\underline{G}_2/R} & \to & 0 \\
 & & & & & & \downarrow & & \\
0 & \to & s^*\Omega^1_{\underline{B}/R} & \to & s^*\Omega^1_{A/R} & \to & s^*\Omega^1_{\underline{G}_1/R} & \to & 0 \\
 & & & & & & \downarrow & & \\
 & & & & & & 0 & & .
\end{array}$$

A diagram chase proves that the arrow α is injective. □

We generalise proposition 2.4. If M is a module over a ring R we write $\mathcal{L}(M)$ for the length of M.

Theorem 2.10: Let R be a discrete valuation ring with quotient field of characteristic 0. Let $\underline{G}$ be a finite group scheme over R, let $\underline{G}'$ be its Cartier dual and n its order. Then

$$\mathcal{L}(s^*\Omega^1_{\underline{G}/R}) + \mathcal{L}(s^*\Omega^1_{\underline{G}'/R}) = \mathcal{L}(R/nR).$$

Proof: We embed $\underline{G}$ into an abelian scheme $\underline{A}$ and define $\underline{B} = \underline{A}/\underline{G}$. Then we have exact sequences

$$0 \to \underline{G} \to \underline{A} \to \underline{B} \to 0$$

$$0 \to \underline{G}' \to \underline{A}' \to \underline{B}' \to 0.$$

Here $\underline{A}'$, $\underline{B}'$ are the dual abelian schemes of $\underline{A}$, $\underline{B}$, see [Oo], [Mu].

We get exact sequences

$$0 \to s^*\Omega^1_{\underline{B}/R} \to s^*\Omega^1_{\underline{A}/R} \to s^*\Omega^1_{\underline{G}/R} \to 0$$

$$0 \to s^*\Omega_{\underline{B}'/R} \to s^*\Omega^1_{\underline{A}'/R} \to s^*\Omega^1_{\underline{G}'/R} \to 0$$

We write g for the dimension $\underline{A}$ or $\underline{B}$. We have

$$s^*\Omega^1_{\underline{G}/R} = \mathrm{Coker}(s^*\Omega^1_{\underline{B}/R} \to s^*\Omega^1_{\underline{A}/R})$$

$$s^*\Omega^1_{\underline{G}'/R} = \mathrm{Coker}(s^*\Omega^1_{\underline{B}'/R} \to s^*\Omega^1_{\underline{A}'/R})$$

$$= \mathrm{Coker}(H^1(\underline{B}, \mathcal{O}_B) \to H^1(\underline{A}, \mathcal{O}_A))' .$$

The last identity uses

$$H^1(\underline{A}, \mathcal{O}_{\underline{A}}) = (s^*\Omega^1_{\underline{A}'/R})', \; H^1(\underline{B}, \mathcal{O}_B) = (s^*\Omega^1_{\underline{B}'/R})'.$$

So we get

$$\mathcal{L}(s^*\Omega^1_{\underline{G}/R}) = \mathcal{L}(\mathrm{Coker}(\Gamma(\underline{B}, \Omega^g_{\underline{B}/R}) \to \Gamma(\underline{A}, \Omega^g_{\underline{A}/R}))$$

$$\mathcal{L}(s^*\Omega^1_{\underline{G}'/R}) = \mathcal{L}(\mathrm{Coker}(H^g(\underline{B}, \mathcal{O}_{\underline{B}}) \to H^g(\underline{A}, \mathcal{O}_A))).$$

This follows by consideration of determinants of the appropriate maps. All maps are here the maps induced from the exact sequences at the beginning. From the Serre-duality theorem we get a commutative diagram

$$\begin{array}{ccccc} H^g(\underline{A}, \mathcal{O}_{\underline{A}}) & \times & \Gamma(\underline{A}, \Omega^g_{\underline{A}/R}) & \to & R \\ \uparrow & & \uparrow & & \uparrow \deg(\varphi) \\ H^g(\underline{B}, \mathcal{O}_{\underline{B}}) & \times & \Gamma(\underline{B}, \Omega^g_{\underline{B}/R}) & \to & R \end{array}$$

But the degree of the isogeny φ coincides with the order of $\underline{G}$. For the notation and for Serre-duality, see [H], chapter III. □

§3. p-divisible groups

Here we discuss the definition and some facts on p-divisible groups. This concept is due to Tate [T].

Definition: R is a noetherian ring, p is a rational prime number and h is a nonnegative integer. A p-divisible group $\underline{G}$ over R of height h is a system

$$\underline{G} = (\underline{G}_k, i_k) \qquad k \geq 0 ,$$

where

(1) each $\underline{G}_k$ is a group scheme of finite order over R. The order of $\underline{G}_k$ is p^{kh}.

(2) for each $k \geq 0$ the sequence of group schemes

$$0 \to \underline{G}_k \xrightarrow{i_k} \underline{G}_{k+1} \xrightarrow{p^k} \underline{G}_{k+1}$$

is exact.

Remarks:

1) The map p^k under (2) is multiplication by p^k. If $\underline{G}$ is any group scheme over $\underline{S}$ and if $n \in \mathbb{N}$ then the composite map

$$\underline{G} \xrightarrow{\text{diag}} \underbrace{\underline{G} \times_{\underline{S}} \cdots\cdots\cdots \times_{\underline{S}} \underline{G}}_{\text{n-times}} \xrightarrow{\mu} \underline{G}$$

is called multiplication by n. If $\underline{G}$ is commutative, it is a homomorphism of group schemes over $\underline{S}$.

2) Let $\underline{G}$ be a finite commutative group scheme over $\underline{S}$ of order n. Then multiplication by n annihilates $\underline{G}$. That

means, there is a commutative diagramm:

$$\begin{array}{ccc} & \nearrow & \underline{S} \\ & & \downarrow s \\ \underline{G} & \xrightarrow{\lambda} & \underline{G} \end{array}$$

where λ is multiplication by n. See [T].

3) If $\underset{\sim}{G} = (\underline{G}_k, i_k)$ is a p-divisible group, we shall prove that the exponent of $\underline{G}_k$ is exactly p^k. The exponent of a finite commutative group scheme $\underline{G} \to \underline{S}$ is the minimal number n such that multiplication by n annihilates $\underline{G}$.

Exactness of the sequence under (2) means that i_k is a closed immersion. Furthermore, i_k has to induce an isomorphism to the kernel of p^k.

Let $\underset{\sim}{G} = (\underline{G}_k, i_k)$, $\underset{\sim}{H} = (\underline{H}_k, j_k)$ be p-divisible groups. A homomorphism

$$\Phi: \underset{\sim}{G} \to \underset{\sim}{H}$$

of p-divisible groups is a system of homomorphisms of group schemes over R

$$\varphi_k: \underline{G}_k \to \underline{H}_k$$

such that the diagrams

$$\begin{array}{ccc} \underline{G}_k & \xrightarrow{\varphi_k} & \underline{H}_k \\ {\scriptstyle i_k}\downarrow & & \downarrow{\scriptstyle j_k} \\ \underline{G}_{k+1} & \xrightarrow{\varphi_{k+1}} & \underline{H}_{k+1} \end{array}$$

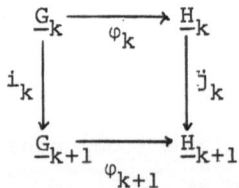

are commutative. A sequence of homomorphisms of p-divisible groups

$$0 \to \underset{\sim}{F} \to \underset{\sim}{G} \to \underset{\sim}{H} \to 0$$

is called exact if the sequences of homomorphisms of group schemes over R

$$0 \to \underline{F}_k \to \underline{G}_k \to \underline{H}_k \to 0$$

are exact in the sense of §1.

We define now for $k,\ell \in R$ with $k \geq 0$, $\ell \geq 1$:

$$i_{k,\ell} = i_{k+\ell-1} \circ \cdots\cdots i_{k+1} \circ i_k .$$

$i_{k,\ell}$ is a closed immersion

$$i_{k,\ell} \colon \underline{G}_k \to \underline{G}_{k+\ell} \quad .$$

We have now

<u>Proposition 3.1</u>: Let $\underset{\sim}{G}$ be a p-divisible group over a noetherian ring R without zero divisors.

(1) The sequences:

$$0 \to \underline{G}_k \xrightarrow{i_{k,\ell}} \underline{G}_{k+\ell} \xrightarrow{p^k} \underline{G}_{k+\ell}$$

are exact for all $k \geq 0$, $\ell \geq 1$.

(2) $\underline{G}_k$ is annihilated by p^k.

(3) There is a homomorphism of group schemes

$$j_{k,\ell} \colon \underline{G}_{k+\ell} \to \underline{G}_\ell$$

such that the following diagram is commutative

$$
\begin{array}{ccc}
\underline{G}_{k+\ell} & \xrightarrow{\;p^k\;} & \underline{G}_{k+\ell} \\
 & \searrow^{j_{k,\ell}} & \uparrow i_{\ell,k} \\
 & & \underline{G}_{\ell}
\end{array}
$$

(4) The sequence of homomorphisms of group schemes

$$0 \to \underline{G}_k \xrightarrow{\;i_{k,\ell}\;} \underline{G}_{k+\ell} \xrightarrow{\;j_{k,\ell}\;} \underline{G}_{\ell} \to 0$$

is exact.

<u>Proof</u>: The $\underline{G}_k$ are all affine schemes:

$$\underline{G}_k = \operatorname{spec} A_k$$

for some R-àlgebra A_k. We write

$$
\begin{array}{lcl}
i_k & : A_{k+1} & \to A_k \\
i_{k,\ell} & : A_{k+\ell} & \to A_k \\
p^k & : A_{\ell} & \to A_{\ell}
\end{array}
$$

for the homomorphisms of R-algebras corresponding to the maps of group schemes with the same name. We also have R-algebra homomorphisms Θ, making the following diagrams commutative:

$$
\begin{array}{ccccc}
 & & A_{k+1} \otimes_R A_{k+1} & \longleftarrow & R \\
 & \swarrow^{\Theta} & \uparrow \psi & & \uparrow s \\
A_k & \xleftarrow{\;i_k\;} & A_{k+1} & \xleftarrow{\;p^k\;} & A_{k+1}\ .
\end{array}
$$

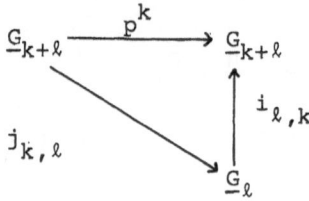

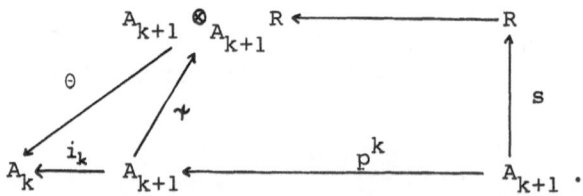

Since the zero-sections are surjective, the maps ψ are surjective. The kernel of ψ is the ideal generated by the image under p^k of the kernel of s.

(1) This is proved by induction on ℓ, the beginning of the induction being obvious. We shall indicate the induction step from $\ell = 1$ to $\ell = 2$. Consider the diagram:

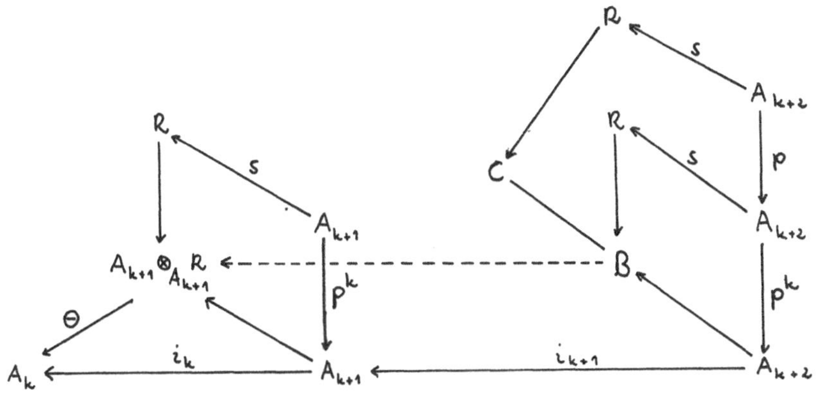

The algebra B is the tensorproduct $A_{k+2} \otimes_{A_{k+2}} R$ formed over the maps p^k and s. Whereas C is the tensorproduct $A_{k+2} \otimes_{A_{k+2}} R$ formed over the maps p^{k+1} and s. The broken line is induced by i_{k+1}. A diagram chase making use of the preliminary remarks shows that this is an isomorphism. The broken line composed with Θ gives the identification of the kernel of p^k with the image of $i_{k,2}$.

(2) Multiplication by p^2 commutes with every R-algebra homomorphism. Consider the diagram

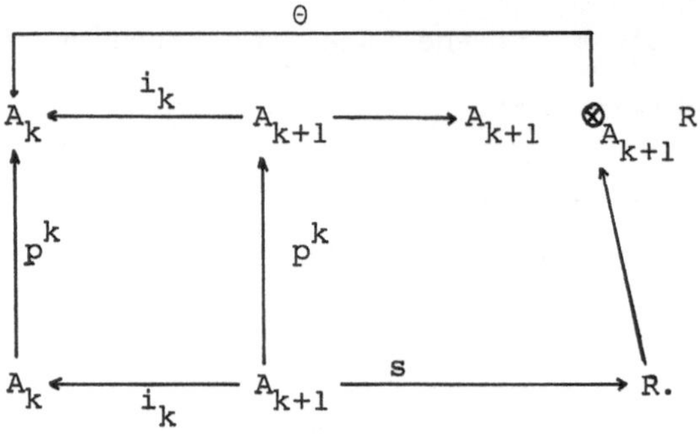

It shows that the map $i_k \circ p^k$ factors through the zero section of A_{k+1}. Hence $p^k \circ i_k$ factors through the zero section of A_{k+1}. Since i_k is surjective, the map

$$p^k : A_k \to A_k$$

factors through the zero section of A_k.

(3) follows from (2).

(4) The problem here is to see that $j_{k,\ell}$ is faithfully flat, everything else is straightforward.

By (1) the kernel of the homomorphism

$$\underline{G}_{k+\ell} \xrightarrow{p^k} \underline{G}_{k+\ell}$$

is a group scheme which is flat over R. From this it follows that p^k is flat, see [M], p. 67. Consider now the diagram:

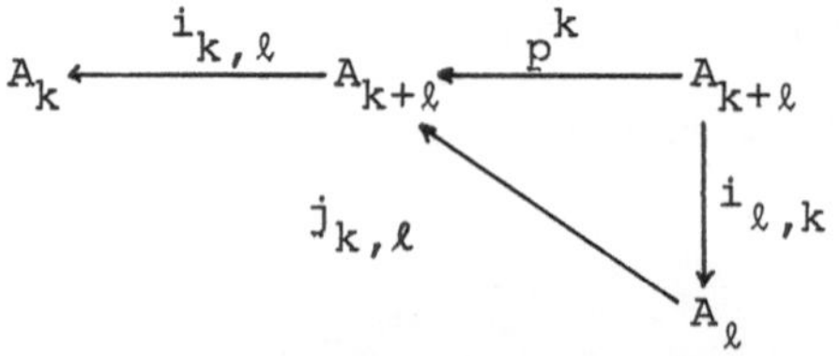

We have already proved that $A_{k+\ell}$ is under $j_{k,\ell}$ a flat A_ℓ-module. We shall prove next that $j_{k,\ell}$ is injective. This is seen by proving

$$\operatorname{rank}_R(p^k(A_{k+\ell})) = p^{h\ell}.$$

To do this tensor the above sequence with the algebraic closure $\hat{K}$ of the quotient field of R. If K has characteristic 0 all the above group schemes become constant and the claim can be checked on the explicit basis for constant group schemes. See §2, and [G,D], II. If the characteristic of K is $p > 0$ then one has to check the claim on the models in [G,D],II.

Now $A_{k,\ell}$ is under $j_{k,\ell}$ a finite ring extension of A_ℓ. Hence by going up, condition (e) of proposition 9 in [Bou] chapter I is satisfied and $A_{k,\ell}$ is a faithfully flat module.

Étale and connected:

In this subsection we assume that R is a complete noetherian local ring with residue field k of characteristic $p > 0$.

Let $\underline{G}$ be a group scheme of finite order over R. Then there is a canonical exact sequence

$$0 \to \underline{G}^0 \to \underline{G} \to \underline{G}^{ét} \to 0 ,$$

where $\underline{G}^0$ is the connected component of 1 in $\underline{G}$ and $\underline{G}^{ét}$ is étale over R. See [G], [Ga] for this. If $\underset{\sim}{G} = (\underline{G}_k, i_k)$ is a p-divisible group over k, then the maps i_k induce maps

$$i_k\colon \underline{G}_k^0 \to \underline{G}_k^0$$
$$i_k\colon \underline{G}_k^{ét} \to \underline{G}_k^{ét} .$$

These can be used to form p-divisible groups $\underset{\sim}{G}^{ét} = (\underline{G}_k^{ét}, i_k)$, $\underset{\sim}{G}^0 = (\underline{G}_k^0, i_k)$. From the sequences

$$0 \to \underline{G}_k^0 \to \underline{G}_k \to \underline{G}_k^{ét} \to 0$$

we get an exact sequence of p-divisible groups

$$0 \to \underset{\sim}{G}^0 \to \underset{\sim}{G} \to \underset{\sim}{G}^{ét} \to 0.$$

We describe here constructions for étale and connected p-divisible groups over R. We start off with connected groups.

Given a natural number n we write

$$\mathcal{A} = R[[x_1,\dots,x_n]]$$

for the ring of formal power series in n variables over R.

Let F be an n-dimensional commutative Lie group over R. F can be described as a system

$$F(x,y) = (f_1(x,y),\dots,f_n(x,y))$$

of n power series in 2n variables which satisfy the following axioms

(i) $F(0,x) = F(x,0) = (x_1,\dots,x_n)$

if $x = (x_1,\dots,x_n)$

(ii) $F(x,F(y,z)) = F(F(x,y),z)$

(iii) $F(x,y) = F(y,x)$

For examples of such see [Ha] chapter 2..

Taking on $\mathcal{A}$ the order topology we have a continuous isomorphism

$$\mathcal{A} \hat{\otimes}_R \mathcal{A} \to R[[x_1,\ldots,x_{2n}]].$$

Using this one sees that there is a unique R-algebra homomorphism

$$\hat{\mu}\colon \mathcal{A} \to \mathcal{A} \hat{\otimes}_R \mathcal{A}$$

satisfying

$$\hat{\mu}(x_i) = f_i(x_1 \hat{\otimes} 1,\ldots,x_n \hat{\otimes} 1;\ 1 \hat{\otimes} x_1,\ldots,1 \hat{\otimes} x_n).$$

Let

$$\alpha(x) = (\alpha_1(x_1,\ldots,x_n),\ldots,\alpha_n(x_1,\ldots,x_n))$$

be the unique n-tuple of power series in n variables satisfying

$$F(x,\alpha(x)) = 0.$$

There is a unique R-algebra homomorphism

$$\hat{i}\colon \mathcal{A} \to \mathcal{A}$$

satisfying

$$\hat{i}(x_i) = \alpha_i(x_1,\ldots,x_n).$$

Define further

$$\hat{s}\colon \mathcal{A} \to R$$

by

$$\hat{s}: x_i \to 0.$$

The R-algebra $\mathcal{A}$ together with the maps $\hat{\mu},\hat{i},\hat{s}$ is a bigebra in the category of continuous R-algebras. That is $\hat{\mu},\hat{i},\hat{s}$ satisfy the commutative diagrams mentioned in §1, only the tensorproducts have to be replaced by their continuous analogs.

We define now inductively

$$\hat{\mu}_n: \mathcal{A} \longrightarrow \underbrace{\mathcal{A} \hat{\otimes}_R \mathcal{A} \hat{\otimes}_R \cdots \hat{\otimes}_R \mathcal{A}}_{n\text{-times}}$$

by $\hat{\mu}_2 = \hat{\mu}$ and

$$\hat{\mu}_{n+1}(x_i) = f_i(\hat{\mu}_n(x_i) \hat{\otimes} 1, 1 \hat{\otimes} x_1, \ldots, 1 \hat{\otimes} x_n).$$

Furthermore put

$$\psi = m \circ \hat{\mu}_p$$

where

$$m: \underbrace{\mathcal{A} \hat{\otimes}_R \mathcal{A} \hat{\otimes}_R \cdots \hat{\otimes}_R \mathcal{A}}_{p\text{-times}} \longrightarrow \mathcal{A}$$

is induced by the continuous multiplication.

$$\psi: \mathcal{A} \to \mathcal{A}$$

is an R-algebra homomorphism corresponding to multiplication by p in the formal Lie group. The following formula is easily seen from the definitions: $\psi^k(x_i) = p^k x_i$ + terms of higher degree. We assume now that ψ is an isogeny, that is $\mathcal{A}$ is under ψ a free $\mathcal{A}$-module of finite rank. The formal group is then said to be divisible. We define a p-divisible group

$$\tilde{F} = (G_k, i_k)$$

as follows:

$$G_k = \operatorname{spec}(\mathcal{A}/_{\langle\psi^k(x_i)\rangle}).$$

Here $\langle\psi^k(x_i)\rangle$ is the ideal in $\mathcal{A}$ generated by the $\psi^k(x_i)$ for $i = 1,\ldots,n$. The bigebra-structure on $\mathcal{A}/_{\langle\psi^k(x_i)\rangle}$ is induced by the maps $\hat{\mu},\hat{i},\hat{s}$. The maps i_k come from the inclusions

$$\langle\psi^k(x_i)\rangle \supseteq \langle\psi^{k+1}(x_i)\rangle.$$

It can be proved by elementary considerations on power series that $\tilde{F}$ is in fact a p-divisible group. Of course each G_k is connected since $\mathcal{A}/_{\langle\psi^k(x_i)\rangle}$ is a local ring.

We have now

Theorem 3.2 (Tate): Let R be a complete noetherian ring whose residue class field has characteristic $p > 0$. Then the map

$$F \longmapsto \tilde{F}$$

is an equivalence between the categories of divisible commutative formal Lie groups over R and the category of connected p-divisible groups over R.

For a proof see [T]. Tate's theorem can now be used to define the dimension of a p-divisible group.

Definition: Let G be a p-divisible group over R with connected component G^o. Let F be an n-dimensional formal group with $\breve{\underset{\sim}{F}} = \underset{\sim}{G}^{O}$. Then n is defined to be the dimension of $\underset{\sim}{G}$.

We shall now give a construction of étale p-divisible groups. Here R is again a complete noetherian ring with residue field k of characteristic $p > 0$. $\hat{k}$ is the separabel algebraic closure of k and $\mathfrak{g}_0$ is the Galois group of $\hat{k}$ over k. Furthermore, let $R_{\acute{e}t}$ be the maximal local étale extension of R. $\mathfrak{g}_0$ lifts to a group of automorphisms of $R_{\acute{e}t}$ over R. We start off with a continuous representation

$$\varphi: \mathfrak{g}_0 \to \mathrm{Aut}((\mathbb{Q}_p/\mathbb{Z}_p)^h) = GL_h(\mathbb{Z}_p)$$

h is a natural number and $\mathbb{Q}_p$, $\mathbb{Z}_p$ are the p-adic number-field and the p-adic integers.

We define now from φ a p-divisible group $\underset{\sim}{\varphi}$. Put

$$\Delta_k = \{u \in (\mathbb{Q}_p/\mathbb{Z}_p)^h \mid p^k \cdot u = 0\} = (\mathbb{Z}/p^k\mathbb{Z})^h .$$

Δ_k is $\mathfrak{g}_0$-invariant. Put

$$A_k = \mathrm{Map}_{\mathfrak{g}_0}(\Delta_k, R_{\acute{e}t})$$

for the ring of $\mathfrak{g}_0$ invariant $R_{\acute{e}t}$ - valued functions on Δ_k. A_k gets a bigebra structure just as the constant group scheme in example 4. The inclusion maps

$$\Delta_k \to \Delta_{k+1}$$

induce R-algebra homomorphisms

$$i_k: A_{k+1} \to A_k.$$

We put

$$\underline{G}_k = \mathrm{spec}(A_k).$$

It is then straightforward to check that

$$\underset{\sim}{\varphi} = (\underline{G}_k, i_k)$$

is a p-divisible group of height h. We have now

Proposition 3.3: Let R be a noetherian local ring. Then the map

$$\varphi \longmapsto \underset{\sim}{\varphi}$$

is an equivalence between the category of continuous representations of $\mathcal{G}_0$ in $GL_n(\mathbb{Z}_p)$ and the category of étale p-divisible groups over R.

This is proved by application of theorem 2.6.

More Examples:

The first example derives from the multiplicative group $\underline{G}_m$. Let p be a prime number then μ_{p^k} is the kernel of the map

$$\underline{G}_m \xrightarrow{p^k} \underline{G}_m.$$

There are obvious inclusions

$$i_k: \mu_{p^k} \to \mu_{p^{k+1}} \quad .$$

The system (μ_{p^k}, i_k) is a p-divisible group of height 1 called

$\underset{\sim}{G}_m(p)$.

Next, let $\underline{A}$ be an abelian scheme of dimension g over R. Assume that the kernel $\underline{A}_k$ of multiplication by p^k on $\underline{A}$ is a flat group scheme over R. This is for example the case if R is a ring of integers in a number field or one of its completetians and R has good reduction modulo $\mathfrak{p}$ for all primes dividing p. The obvious inclusions $i_k: \underline{A}_k \to \underline{A}_{k+1}$ make

$$\underline{A}(p) = (\underline{A}_k, i_k)$$

into a p-divisible group of height 2g over R.

Let E be an elliptic curve over $\mathbb{Z}_p$ that has good reduction modulo p. It is interesting to consider the decomposition of $\underline{E}(p)$ into its connected and étale parts. One finds:

$\underline{E}(p)$ is connected $\Longleftrightarrow$ the Hasse-invariant of E is O.

In case the Hasse-invariant of E is not zero one has an exact sequence

$$0 \to \underline{E}(p)^0 \to \underline{E}(p) \to \underline{E}(p)^{ét} \to 0$$

where $E(p)^0$ is a connected p-divisible group of height 1. See [Se] for this.

<u>Modules of differentials</u>:

We now use Tate's theorem to compute the modules of differentials of the constituents of a p-divisible group.

<u>Proposition 3.4</u>: Let R be a noetherian local ring with

residue class field of characteristic $p > 0$. Let $\underset{\sim}{G} = (\underline{G}_k, i_k)$ be an n-dimensional p-divisible group over R. Then

$$s^*\Omega^1_{\underline{G}_k/R} = ({}^R/_{p^k R})^n.$$

Proof: The differential module of an étale group is zero. So, using proposition 1.1 we may assume that G is connected. By theorem 3.2 we may choose a divisible n-dimensional formal Lie group F with $\underset{\sim}{\tilde{F}} = \underset{\sim}{G}$. Let $\mathcal{A} = R[[x_1, \ldots, x_n]]$ be the ring of formal power series over R and let $\psi, \hat{\mu}, \hat{i}, \hat{s}$ be as defined before theorem 3.2. The module of formal differentials

$$\hat{\Omega}^1_{\mathcal{A}/R}$$

is a free module of rank n over $\mathcal{A}$:

$$\hat{\Omega}^1_{\mathcal{A}/R} = \mathcal{A}\, dx_1 \oplus \ldots \oplus \mathcal{A} \cdot dx_n.$$

The derivation being

$$Df = \frac{\partial f}{\partial x_1} dx_1 + \ldots + \frac{\partial f}{\partial x_n} dx_n.$$

From the formula

(*) $\quad \psi^k(x_i) = p^k x_i$ + terms of higher degree

we get: $\quad D\psi^k(x_i) = p^k dx_i$ + nonconstant terms.

The homomorphism

$$\mathcal{A} \longrightarrow \mathcal{A}/_{<\psi^k(x_i)>} = A_k$$

is continuous: [T]. Hence we get a map

$$\hat{\Omega}^1_{\mathcal{A}/R} \to \hat{\Omega}^1_{A_k/R} = \Omega^1_{A_k/R} = \Omega^1_{\underline{G}_k/R} .$$

By [Gr] we get

$$\Omega^1_{A_k/R} = \bigoplus_{i=1}^{n} \mathcal{A} / \langle \psi^k(x_i), \frac{\partial f}{\partial x_i} \psi^k(x_1), \ldots, \frac{\partial f}{\partial x_i} \psi^k(x_n) \rangle$$

Using (✱) we find the required formula.

The Tate module:

We assume here that R is a complete discrete valuation ring with quotient field K and residue field k. We assume that $\operatorname{char} K = 0$ and $\operatorname{char} k = p > 0$. $\hat{K}, \hat{k}$ are the separable algabraic closures of K and k. $\mathfrak{g}$ is the galoisgroup of $\hat{K}$ over K. Let $\underset{\sim}{G} = (\underline{G}_k, i_k)$ a p-divisible group of height h over R. Then we have maps

$$j_{j,1} \colon \underline{G}_{k+1} \to \underline{G}_k .$$

These induce maps

$$j_k \colon \underline{G}_{k+1}(\hat{K}) \to \underline{G}_k(\hat{K}) .$$

The limit $T(\underset{\sim}{G}) = \lim\limits_{k} \underline{G}_k(\hat{K})$ is called the Tate module of G.

Since the $\underline{G}_k \otimes_R \hat{K}$ are étale and hence constant the group $T(G)$ gets a natural $\mathbb{Z}_p$-module structure. As $\mathbb{Z}_p$-module we have $T(\underset{\sim}{G}) = \mathbb{Z}_p^n$. The galoisgroup $\mathfrak{g}$ acts continuously on $T(\underset{\sim}{G})$. We shall describe examples of this action in §5. If

$\varphi\colon \underline{G} \to \underline{H}$ is a homomorphism of p-divisible groups we get an induced homomorphism

$$T(\varphi)\colon T(\underline{G}) \to T(\underline{H}).$$

Clearly the image of $T(\varphi)$ is a $\mathbb{Z}_p$-direct summand of $T(\underline{H})$. It is also $\mathfrak{g}$ invariant.

Theorem 3.5: Let $\underline{H}$ be a p-divisible group over R. Let furthermore $M \leq T(\underline{H})$ be a $\mathfrak{g}$-invariant $\mathbb{Z}_p$-direct summand. Then there is a p-divisible group $\underline{G}$ over R and a homomorphism of p-divisible groups $\varphi\colon \underline{G} \to \underline{H}$ such that φ induces an isomorphism

$$T(\varphi)\colon T(\underline{G}) \xrightarrow{\sim} M \subseteq T(\underline{H}).$$

A proof of this is contained in section 4.2 of [T].

Remark: For the application in [Sch] note that a $\mathbb{Q}_p$-subspace

$$W \subseteq \mathbb{Q}_p \otimes_{\mathbb{Z}_p} T(\underline{H})$$

intersects $T(\underline{H})$ in a $\mathbb{Z}_p$-direct summand. In general one has then to go to an extension so that this summand gets galois-invariant.

§4 A theorem of Raynaud

Here R is a complete discrete valuation ring with quotient field K and residue field k. We assume that char K = 0 and char k = p > 0. $\mathfrak{g}$, $\mathfrak{g}_0$ are the galoisgroups

$$\mathfrak{g} = \mathfrak{g}al(\hat{K}:K), \quad \mathfrak{g}_0 = \mathfrak{g}al(\hat{k}:k)$$

where $\hat{K},\hat{k}$ are the separable algebraic closures of K and k.

Let $\underline{G}$ be a commutative group scheme of finite order over R which is annihilated by multiplication by p. Raynaud calls these group schemes of type (p,...,p). The scheme $\underline{G}$ is affine, $\underline{G}$ = spec(A) for some R-algebra A. A is a free R-module of rank p^r. The group scheme

$$\underline{G} \otimes_R K = \underline{G} \times_{\mathrm{spec}(R)} \mathrm{spec}(K)$$

is reduced, since K is of characteristic 0, see [Ca], page 109. So $A \otimes_R K$ is a product of finite extensions of K, and $\underline{G} \otimes_R K$ is étale over K. The ring R is some order in a product of finite extension of K. The order of $\underline{G}$ is a power of p. This follows from the general structure theorem on étale finite groups [G,D], II §5.

From this it follows that the group of $\hat{K}$-valued points of $\underline{G}$

$$\underline{G}(\hat{K}) = \underline{G}(\mathrm{spec}(\hat{K}))$$

is isomorphic to

$$\underline{G}(\hat{K}) = (\mathbb{F}_p)^r.$$

Multiplication by natural numbers makes $G(\hat{K})$ into an $\mathbb{F}_p$ - vectorspace of dimension r. Hence the galoisgroup $\mathfrak{g}$ acts linearly on $\underline{G}(\hat{K})$. We write

$$\rho_{\underline{G}}: \mathcal{G} \longrightarrow GL_r(\mathbb{F}_p)$$

for the corresponding representation. We define

$$\chi_{\underline{G}}: \mathcal{G} \longrightarrow (\mathbb{F}_p)^* = GL(\Lambda^r(\mathbb{F}_p^r))$$

as the determinant representation of $\rho_{\underline{G}}$. We shall be interested in the representation $\chi_{\underline{G}}$. To analyse $\chi_{\underline{G}}$ we have to introduce the following: K_{nr} = maximal unramified extension of K

K_t = maximal tamely ramified extension of K.

R_{nr} = integral closure of R in K_{nr}.

R_t = integral closure of R in K_t.

$I = \mathcal{G}al(\hat{K}:K_{nr}) \subseteq \mathcal{G}$

$I_p = \mathcal{G}al(\hat{K}:K_t) \subseteq \mathcal{G}$

$I_t = I/_{I_p} = \mathcal{G}al(K_t:K_{nr})$

The notation here is the same as in [Se], §1. ν is the valuation of K and $e = \nu(p)$. We say that R is strictly henselian if R has no étale local extension rings. This means that $K_{nr} = K$.

The galoisgroup $\mathcal{G}$ acts on the groups of p-th roots of unity in $\hat{K}$. This defines a homomorphism

$$\overline{\chi}_0: \mathcal{G} \to \mathrm{Aut}(\mu_p(\hat{K})) = (\mathbb{F}_p)^*.$$

The group of (p-1)-th roots of unity $\mu_{p-1}(\hat{K})$ is contained in R. Applying the residue map we get an isomorphism

$$\mu_{p-1}(\hat{K}) \xrightarrow{\sim} \mathbb{F}_p^*.$$

Let π be a uniformising element for R. The field

$$K(\sqrt[p-1]{\pi})$$

is a tamely ramified Galois extension of k. For $g \in \mathfrak{g}$ we have

$$g(\sqrt[p-1]{\pi}) = \sqrt[p-1]{\pi} \cdot \zeta$$

with a $\zeta \in \mu_{p-1}(\hat{K})$. Using the above isomorphism $\mu_{p-1}(\hat{K}) \cong \mathbb{F}_p^*$ we may extend the map $g \to \zeta$ to a homomorphism

$$\tau_p : \mathfrak{g} \to \mathbb{F}_p^* .$$

Note that both τ_p and $\bar{\chi}_0$ have to vanish on the pro-p-group I_p. We have

$$\bar{\chi}_0 = \tau_p^e$$

on I. See [Se] for this fact.

If M is a finitely generated R-torsion module we define

$$\mathcal{L}(M)$$

to be the length of the module M. The length has the following properties:

1) $\mathcal{L}(R/_{aR}) = \nu(a)$ for any $a \in R$

2) If $0 \to M_1 \to M_2 \to M_3 \to 0$ is an exact sequence of finitely generated R-torsion modules then

$$\mathcal{L}(M_2) = \mathcal{L}(M_1) \cdot \mathcal{L}(M_3).$$

<u>Examples</u>:

We shall describe here for various examples the character $\chi_{\underline{G}}$.

Example 1: $\underline{G} = \mu_p$. Here $\underline{G} = \mathrm{spec}(A)$ where $A = R[t]/_{<t^p-1>}$.

If K contains a primitive p-th root of unity then $A \otimes_R K$ is a product of p copies of K. If K doesn't contain a primitive p-th root of unity $A \otimes_R K$ is the product of K with the field $K[t]/_{<t^{p-1} + \ldots + 1>}$.

Giving a $\hat{K}$ valued point of $\underline{G}$ amounts to selecting a p-th root of unity for t. We get

$$\bar{\chi}_0 = \rho_{\mu_p} = \chi_{\mu_p} .$$

Example 2: Étale groups.

We start off with a continuous representation

$$\varphi\colon \mathfrak{G}_0 \to GL_r(\mathbb{F}_p).$$

The group $\mathfrak{G}_0 = \mathfrak{G}al(\hat{k}:k)$ can be identified with $\mathfrak{G}al(K_{nr}:K)$, so we may consider the ring of $\mathfrak{G}_0$- invariant functions

$$A = Map_{\mathfrak{G}_0}(\mathbb{F}_p^r, R_{nr}) .$$

Define the bigebra structure on A by the same formulas as in case of the constant group scheme of §1 example 4. This defines an étale group scheme $\underline{\varphi} = spec(A)$ of type $(p,\ldots,p)$ over R. In our case R_{nr} coincides with the maximal étale extension of R. It follows from [G,D] II, §5 that every étale group scheme of type $(p,\ldots,p)$ is of the form $\underline{\varphi}$ for some representation φ. We assert now that $\rho_{\underline{\varphi}} = \tilde{\varphi}$, where $\tilde{\varphi}$ is φ composed with the projection $\mathfrak{G} \to \mathfrak{G}_0$. Note that $\rho_{\underline{\varphi}}$ is trivial on the ramification group I_p.

Example 3: The groups $\underline{G}_{a,b}$

Let $a,b \in R$ be two elements of R with $a \cdot b = p$. We have defined the groups $\underline{G}_{a,b}$ in §2. $\underline{G}_{a,b}$ is of order p and hence annihilated by p. The galois representation $\rho_{\underline{G}_{a,b}}$ can now be described as follows. The field K contains the $(p-1)$-th roots of unity $\mu_{p-1}(\hat{K})$. From the residue map we have an isomorphism

$$\chi: \mu_{p-1}(K) \xrightarrow{\sim} \mathbb{F}_p^*.$$

For any $g \in \mathfrak{g}$ we may write

$$g(\sqrt[p-1]{a}) = \sqrt[p-1]{a} \cdot \zeta$$

with a $(p-1)$-th root of unity ζ. We define the Kummer character

$$\chi_a: \mathfrak{g} \to \mathbb{F}_p^*$$

as $\chi_a(g) = \chi(\zeta)$. It is an interesting exercise using the explicit formulas for the multiplication in $\underline{G}_{a,b}$ to prove

$$\rho_{\underline{G}_{a,b}} = \chi_{\underline{G}_{a,b}} = \chi_a.$$

Assume for a moment that R is strictly henselian. Writing $a = \pi^{\nu(a)} \cdot u$ with a unit u, we find that $\chi_a = (\tau_p)^{\nu(a)}$. From proposition 2.3 we have $\nu(a) = \mathcal{L}(s^*\Omega^1_{\underline{G}_{a,b}/R})$. This is a special case of theorem 4.5.

<u>Group schemes with</u> F_q^*<u>-action</u>:

In addition to the previous assumptions we assume in this subsection that R is strictly henselian. Otherwise we use the same notation. Let $q = p^r$ for some natural number r. Since R is strictly henselian it contains the group of $(q-1)$-th roots of unity.

Let now $\underline{G} = \mathrm{spec}(A)$ be a commutative group scheme of finite order over R. An $\mathbb{F}_q$-compatible system of endomorphisms of $\underline{G}$ is a map

$$[\quad] : \mathbb{F}_q \to \mathrm{End}_{\mathrm{bialg}}(A)$$

such that

$$[1] = \mathrm{id}$$

$$[a]\circ[b] = [ab]$$

$$m\circ([a] \otimes [b])\circ\mu = [a+b]$$

for all $a,b \in \mathbb{F}_q$. Here $m: A \otimes_R A \to A$ is the multiplication of A. . For example, the multiplications by a natural number define an $\mathbb{F}_p$-compatible system of endomorphisms on any commutative group scheme.

<u>Definition</u>: A group scheme $\underline{G}$ with an $\mathbb{F}_q^*$-action is a commutative group scheme of order q together with an $\mathbb{F}_q$-compatible system of endomorphisms. Note that $\underline{G}$ is then already annihilated by p.

Given a group scheme with an $\mathbb{F}_q^*$-action one can decompose the augmentation ideal, that is the kernel of ε, of A according to the orthonogal idempotents:

$$e_\chi = \frac{1}{q-1} \sum_{\lambda\in\mathbb{F}_q^*} \chi^{-1}(\lambda)[\lambda].$$

Here $\chi: \mathbb{F}_q^* \to \mu_{q-1}(\hat{K}) \subseteq R$ is a character of $\mathbb{F}_q^*$. One gets as in [O] :

<u>Proposition 4.1</u> (Raynaud): Let $\underline{G} = \mathrm{spec}(A)$ be a group scheme over R with an $\mathbb{F}_q^*$-action for $q = p^r$. Then there are elements $\delta_1,\dots,\delta_r$ such that

$$A = {}^{R[x_1,\dots,x_r]}\big/_{\langle x_1^p-\delta_1 x_2,\,x_2^p-\delta_2 x_3,\dots,x_r^p-\delta_r x_1\rangle} .$$

We call $\delta_1,\dots,\delta_r$ the parameters of $\underline{G}$. A proof is contained in [R], mind the assumptions on R that we have made here. In [R] one also finds a formula for the comultiplication μ of A. The above formula can now be used to prove

<u>Proposition 4.2</u>: Let $\underline{G} = \mathrm{spec}(A)$ be a group scheme over R with an $\mathbb{F}_q^*$-action for $q = p^r$. Let $\delta_1,\dots,\delta_r$ be the parameters of $\underline{G}$. Then

1) $s^*\Omega^1_{\underline{G}/R} \cong {}^R/_{\delta_1 R} \oplus \dots \oplus {}^R/_{\delta_r R}$

2) $\mathcal{L}(s^*\Omega^1_{\underline{G}/R}) = \nu(\delta_1) + \dots + \nu(\delta_r)$

<u>Proof</u>: We have for $B = R[x_1,\dots,x_r]$

$$\Omega^1_{B/R} = B\cdot dx_1 \oplus \dots \oplus B\cdot dx_r ,$$

where the derivation is

$$D\colon B \to \Omega^1_{B/R}$$
$$Df = \frac{\partial f}{\partial x_1}\,dx_1 + \dots + \frac{\partial f}{\partial x_r}\,dx_r .$$

Let $A = {}^B\big/_{\langle x_1^p-\delta_1 x_2,\dots,x_r^p-\delta_r x_1\rangle}$ then

$\Omega^1_{A/R}$ is $\Omega^1_{B/R}$ divided by the submodule generated by the

$$D(x_i^p-\delta_i x_{i+1}) .$$ □

We can now prove the result

Proposition 4.3: Let $\underline{G} = \mathrm{spec}(A)$ be a group scheme over R with an $\mathbb{F}_q^*$- action for $q = p^r$. Let

$$d = \mathcal{L}(s^*\Omega^1_{\underline{G}/R}),$$

then

$$\chi_{\underline{G}} = (\tau_p)^d.$$

Proof: Let $\delta_1,\ldots,\delta_r$ be the parameters of $\underline{G}$. Then one finds easily that

$$\chi_{\underline{G}} = (\tau_p)^d$$

with $d = \nu(\delta_1) + \ldots + \nu(\delta_r)$. See [R] section 3.4. By proposition 4.2 we have

$$d = \mathcal{L}(s^*\Omega^1_{\underline{G}/R}). \qquad \square$$

Generalization:

In this subsection R is again a strictly henselian local ring of unequal characteristic. Otherwise the notations from the beginning of this chapter are valid. We quote from [R], Corollary 3.3.7.

Theorem 4.4: Let R be a strictly henselian ring with $e \leq p-1$. Let $\underline{G}$ be a group scheme over R which is commutative, of finite order and annihilated by a power of p. Then $\underline{G}$ has a decomposition series $\underline{G}_i$, $i = 0,\ldots,k$ such that

$$\underline{G}_{i+1}/\underline{G}_i$$

has an $\mathbb{F}_q^*$-action for some $q = p^r$.

Remark: A decomposition series of $\underline{G}$ is a sequence $\underline{G}_0,\ldots,\underline{G}_k$ of group schemes with $\underline{G}_k = \underline{G}$ and $\underline{G}_0 \mathrel{\hat{=}} \mathrm{spec}(R)$, together with closed immersions

$$0 \to \underline{G}_i \to \underline{G}_{i+1}.$$

By [G], [Ra] the faithfully flat quotient $\underline{G}_{i+1}/\underline{G}_i$ exists.

Theorem 4.5: Let R be a strictly henselian ring with $e \leq p-1$. Let $\underline{G}$ be a group scheme over R which is commutative, of finite order and annihilated by p. Let

$$\mathcal{L}(s^*\Omega^1_{\underline{G}/R}) = d.$$

Then

$$\chi_{\underline{G}} = (\tau_p)^d.$$

Proof: We use here the exactness of $s^*\Omega^1$ from theorem 2.9 together with the multiplicativity of $\mathcal{L}$. The result then follows by an obvious induction argument along a decomposition series from theorem 4.4. Note that if

$$0 \to \underline{G}_1 \to \underline{G}_2 \to \underline{G}_3 \to 0$$

is exact, then the Galois modules $\underline{G}_i(\hat{K})$ satisfy

$$\underline{G}_2(\hat{K}) \cong \underline{G}_1(\hat{K}) \times \underline{G}_3(\hat{K}).$$

Remark: This is more or less theorem 4.11 from [R].

Globalization:

We apply theorem 4.5 to a global situation. We fix the following notations

K is a finite extension field of $\mathbb{Q}$ of degree m.

$\mathcal{O}$ is its ring of integers.

K_v is the completion of K at the place v.

$\mathcal{O}_v$ is the ring of integers in K_v.

p is a prime number and K is assumed to be unramified at p.

$v_1,\dots,v_r$ are the places extending p.

m_i is the degree of the extension $\mathbb{Q}_p \subseteq K_{v_i}$.

$\hat{\mathbb{Q}} = \hat{K} \subseteq \hat{\mathbb{Q}}_p = \hat{K}_v$ are the algebraic closures of the various fields.

$\mathcal{G}_K \subseteq \mathcal{G}_{\mathbb{Q}}$ are the absolute Galois groups of K and $\mathbb{Q}$.

Given a representation ρ of a group G on a $\mathbb{F}_p$-vectorspace and a subgroup $H \subseteq G$ we write $\rho|_H$ for the restriction of ρ to H. If H is of finite index in G we denote by $\mathrm{Ind}_H^G(\rho)$ the induction of a representation of H to G. Given two characters $\chi_1, \chi_2 \colon G \to \mathbb{F}_p^*$ we write $\chi_1 \otimes \chi_2$ for their tensorproduct.

$\mathcal{G}_p$ is the decomposition group at p; $\mathcal{G}_p \subseteq \mathcal{G}_{\mathbb{Q}}$.

I_p is the ramification group at p; $I_p \subseteq \mathcal{G}_p$

$\mathcal{G}_1,\dots,\mathcal{G}_r$ are the decomposition groups at $v_1,\dots,v_r$; $\mathcal{G}_i \subseteq \mathcal{G}_K$.

$I_1,\dots,I_r$ are the ramification groups at $v_1,\dots,v_r$; $I_i \subseteq \mathcal{G}_i$.

$\varepsilon \colon \mathcal{G}_{\mathbb{Q}} \to \mathbb{F}_p^*$ is the determinant character of the permutation representation of $\mathcal{G}_{\mathbb{Q}}$ on $\mathcal{G}_{\mathbb{Q}}/\mathcal{G}_K$.

Given a character $\chi \colon \mathcal{G}_K \to \mathbb{F}_p^*$ we define

$$\chi^* = \Lambda^m(\mathrm{Ind}_{\mathcal{G}_K}^{\mathcal{G}_{\mathbb{Q}}}(\chi))$$

for the determinant character of the induced representation.

Given a finite group scheme $\underline{G}$ over $\mathcal{O}$ which is commutative and annihilated by multiplication by p we again have

$$\underline{G}(\hat{K}) = (\mathbb{F}_p)^t$$

for some t. The Galois group $\mathcal{G}_K$ acts linearly on $\underline{G}(\hat{K})$. We denote this representation again by $\rho_{\underline{G}}$ and its determinant representation by $\chi_{\underline{G}}$. Similarly, we have the representations $\rho_{\underline{G}_i}$ and $\chi_{\underline{G}_i}$ if $\underline{G}_i = \underline{G} \otimes_{\mathcal{O}} \mathcal{O}_{v_i}$. Identifying $\underline{G}_i(\hat{K}_{v_i})$ with $\underline{G}(\hat{K})$, we have

$$\rho_{\underline{G}}|\mathcal{G}_i = \rho_{\underline{G}_i} \qquad \chi_{\underline{G}}|\mathcal{G}_i = \chi_{\underline{G}_i}.$$

$\mathcal{G}_i$ is here identified with the absolute Galois group of K_{v_i}. $\bar{\chi}_0$ is as before the cyclotomic character $\bar{\chi}_0 = \chi_{\mu_p}$.

<u>Theorem 4.6</u>: Let $\underline{G}$ be a finite commutative group scheme over $\mathcal{O}$ annihilated by p. Assume that each $\underline{G}_i = \underline{G} \otimes_{\mathcal{O}} \mathcal{O}_{v_i}$ is flat over $\mathcal{O}_{v_i}$. We then have

$$p^d = \#(s^*\Omega^1_{\underline{G}/\mathcal{O}})$$

for some nonnegative integer d. The character

$$\chi^*_{\underline{G}} \otimes \varepsilon^t \otimes \chi_0^{-d}: \mathcal{G}_{\mathbb{Q}} \to \mathbb{F}_p^*$$

is unramified at p, that is, it is trivial on I_p.

<u>Proof</u>: The group $s^*\Omega^1_{\underline{G}/\mathcal{O}}$ is annihilated by p so its order is a power of p. This settles the first claim. By the base change isomorphism we have

$$s^*\Omega^1_{\underline{G}_i/\mathcal{O}_{v_i}} = s^*\Omega^1_{\underline{G}/\mathcal{O}} \otimes_{\mathcal{O}} \mathcal{O}_{v_i}.$$

Putting

$$d_i = \mathcal{L}(s^*\Omega^1_{G_i/\mathcal{O}_{v_i}})$$

we have

$$d = \sum_{i=1}^{r} m_i\, d_i.$$

Let $\tilde{\mathcal{O}}_i$ be the ring of integers in the maximal unramified extension $(K_{v_i})_{nr}$ of K_{v_i}. We have a natural identification

$$I_{v_i} = \mathcal{G}al(\hat{K}_{v_i} : (K_{v_i})_{nr}).$$

The ring $\tilde{\mathcal{O}}_i$ is strictly henselian. We have

$$d_i = \mathcal{L}(s^*\Omega^1_{\underline{G}} \otimes_{\mathcal{O}} \tilde{\mathcal{O}}_{v_i}|\tilde{\mathcal{O}}_{v_i}).$$

Since

$$\tilde{\underline{G}}_i = \underline{G}_i \otimes_{\mathcal{O}_{v_i}} \tilde{\mathcal{O}}_{v_i}$$

is by assumption flat over $\tilde{\mathcal{O}}_{v_i}$ we may apply theorem 4.5 and we get:

$$1 = \chi_{\tilde{\underline{G}}_i} \otimes \bar{\chi}_0^{-d_i} = (\chi_{\underline{G}} \otimes \bar{\chi}_0^{-d_i})|_{I_{v_i}}.$$

We may apply theorem 4.5 since the valuation of p in the local rings $\tilde{\mathcal{O}}_{v_i}$ is always 1. This also has $\bar{\chi}_0 = \tau_p$ as consequence.

The following is a standard identity from the representation theory of groups (see [Ser])

$$(\mathrm{Ind}_{\mathcal{G}_K}^{\mathcal{G}_{\mathbb{Q}}}(\rho_{\underline{G}}))|_{I_p} = \bigoplus_{i=1}^{r} \mathrm{Ind}_{I_{v_i}}^{I_p}(\rho_{\underline{G}}|_{I_{v_i}})$$

The result now follows by taking determinants of both sides. □

Remarks: 1) Theorem 4.6 is applied in [Wü] in the situation where one already knows (from stable reduction) that the character

$$\chi^*_{\underline{G}} \otimes \varepsilon^t \otimes \chi_0^{-d}$$

is unramified at all primes different from p. Then it has to be trivial by class field theory. In the application $\underline{G}$ is the kernel of an isogeny between abelian schemes having good reduction at all places extending p. From this follows that G is flat at these places.

2) If $\underline{G}$ is already flat (of finite order) over $\mathcal{O}$ then the same argument shows that the above character is trivial.

§5 A theorem of Tate

Here we discuss a theorem of Tate on the action of the Galois group on the Tate module of a p-divisible group. R is again a complete discrete valuation ring of unequal characteristic. K is the quotient field of R and k is the residue field of R. $\mathfrak{G}$, $\mathfrak{G}_0$ are the Galois groups of the separable algebraic closures $\hat{K}$, $\hat{k}$ over K and k respectively. Let

$$\underset{\sim}{G} = (\underset{\sim}{G}_k, i_k)$$

be a p-divisible group of height h over R. Then the Galois group $\mathfrak{G}$ acts $\mathbb{Z}_p$-linearly on the Tate module $T(\underset{\sim}{G})$ of $\underset{\sim}{G}$. We call this representation $\rho_{\underset{\sim}{G}}$:

$$\rho_{\underset{\sim}{G}} : \mathfrak{G} \to \mathrm{Aut}(T(\underset{\sim}{G})) = GL_h(\mathbb{Z}_p)$$

The corresponding determinant character is called $\chi_{\underset{\sim}{G}}$:

$$\chi_{\underset{\sim}{G}}: \mathfrak{G} \to \mathrm{Aut}(\Lambda^h(T(\underset{\sim}{G}))) = \mathbb{Z}_p^*$$

Examples:

We shall describe the character $\chi_{\underset{\sim}{G}}$ for two examples.

Example 1: $\underset{\sim}{G}_m(p)$.

The p-divisible group $\underset{\sim}{G}_m(p)$ has height 1 and dimension 1 and

$$\chi_0 := \chi_{\underset{\sim}{G}_m(p)} = \rho_{\underset{\sim}{G}_m(p)}$$

is called the cyclotomic character of $\mathfrak{G}$. Let $\mathbb{Z}_p^* \to \mathbb{F}_p^*$ be the canonical quotient map. Following χ_0 by this map we get a character of $\mathfrak{G}$ with values in $\mathbb{F}_p^*$, it coincides with the character $\overline{\chi}_0 = \chi_{\mu_p}$ defined in §4.

Example 2: Étale groups:

Given a continuous representation

$$\varphi: \mathfrak{g}_0 \to \mathrm{Aut}((\mathbb{Q}_p/\mathbb{Z}_p)^h) = GL_h(\mathbb{Z}_p)$$

we have defined a p-divisible group $\underset{\sim}{\varphi}$ in §3. $\underset{\sim}{\varphi}$ has height h and dimension 0. We have a canonical map $\mathfrak{g} \to \mathfrak{g}_0$, so by composition φ defines a homomorphism

$$\tilde{\varphi}: \mathfrak{g} \to GL_h(\mathbb{Z}_p) .$$

It can be checked easily that

$$\tilde{\varphi} = \rho_{\underset{\sim}{\varphi}}$$

and

$$\chi_{\underset{\sim}{\varphi}} = \Lambda^h \tilde{\varphi}.$$

Note that both $\rho_{\underset{\sim}{\varphi}}$ and $\chi_{\underset{\sim}{\varphi}}$ vanish on the ramification group, that is on the kernel of $\mathfrak{g} \to \mathfrak{g}_0$.

<u>Tate's theorem</u>:

Let C be the completion of the algebraic closure $\hat{K}$ of K. The Galois group $\mathfrak{g}$ acts continuously on $\hat{K}$, hence this action extends to an action on C:

$$\mathfrak{g} \to \mathrm{Aut}(C).$$

The p-adic integers are naturally embedded in R hence in C. So we may for any character $\psi: \mathfrak{g} \to \mathbb{Z}_p^*$ define the following action of $\mathfrak{g}$ on C:

$$\sigma(\lambda) = \psi(\sigma) \cdot \sigma(\lambda)$$

This module for the group $\mathfrak{g}$ is denoted by $C(\psi)$ and is called the Tate twist of the Galois module C by ψ. For an integer t we also introduce the notation

$$C(\chi_0^t) =: C(t)$$

Given a p-divisible group G, Tate describes in [T] the structure of the Galois module

$$T(G) \otimes_{\mathbb{Z}_p} C.$$

In the application, [Sch], we need only information on the determinant-character $\chi_{\underset{\sim}{G}}$. We have

<u>Theorem 5.1</u>: Assume that R is a strictly henselian complete discrete valuation ring with quotient field of characteristic 0 and residue field of characteristic p. Let $\underset{\sim}{G}$ be a p-divisible group of height h and dimension d over R. Then

$$\chi_{\underset{\sim}{G}} = \chi_0^d .$$

This formulation is due to Raynaud, a proof is contained in [R]. First of all, the p-divisible group $\underset{\sim}{G}$ can be supposed to be connected since both the dimension and the determinant character $\chi_{\underset{\sim}{G}}$ coincide for $\underset{\sim}{G}$ and its connected component. Then $\underset{\sim}{G}$ comes, as is explained in §3, from a formal group F. Raynaud then uses the deformation theory of formal groups together with a purity argument to prove the result. Another formulation is

<u>Theorem 5.2</u>: Let R be a complete discrete valuation

ring with quotient field of characteristic 0 and residue characteristic p. Let $\underset{\sim}{G}$ be a p-divisible group of dimension d and height h over R. Then there is an isomorphism of $\mathcal{G}$-modules:

$$\Lambda^h(T(G)) \otimes_{\mathbb{Z}_p} C \cong C(d).$$

<u>Proof</u>: Let $\hat{K}$ be the algebraic closure of K and K_{nr} the maximal unramified extension of K in $\hat{K}$. R_{nr} is the integral closure of R in K_{nr}. R_{nr} is a complete discrete valuation ring, it is strictly henselian. So, we may apply theorem 5.1 to the p-divisible group

$$\underset{\sim}{G} \otimes_R R_{nr}.$$

Let I be the absolute Galois group of K_{nr}. I is a normal subgroup of $\mathcal{G}$. We have a natural identification:

$$\mathcal{G}/_I \cong \mathcal{G}al(\hat{k}:k)$$

where k is the residue field of R and $\hat{k}$ its algebraic closure. Clearly we have

$$\chi_{\underset{\sim}{G} \otimes_R R_{nr}} = \chi_{\underset{\sim}{G}}|_I .$$

$|_I$ denotes the restriction of $\chi_{\underset{\sim}{G}}$ to I. By application of theorem 5.1 we find that

$$\chi_{\underset{\sim}{G}}|_I = \chi_0^d|_I.$$

There is a character $\Theta: \mathcal{G} \to \mathbb{Z}_p$ which is trivial on I and satisfies

$$\chi_{\underset{\sim}{G}} \cdot \Theta = \chi_0^d .$$

Θ has to have a finite image in $\mathbb{Z}_p^*$. Its image has to be then in the (p-1)-th roots of unity in $\mathbb{Z}_p$. Let L by the fixed field of the kernel of Θ. By Kummer-theory, [C,F], there is an element $a \in L$ with $\sigma(a) = \Theta(\sigma)\cdot a$ for all $\sigma \in \mathfrak{G}$. Define now

$$\varphi: C(\chi_{\underset{\sim}{G}}\cdot\Theta) \to C(\chi_o)$$

by

$$\varphi: c \to a\cdot c$$

φ is an isomorphism of Galois modules, as is seen by the following computation:

$$\begin{aligned}\varphi(\sigma(c)) &= \varphi(\chi_{\underset{\sim}{G}}\cdot\Theta(\sigma)\cdot\sigma(c))\\ &= a\cdot\Theta(\sigma)\cdot\chi_{\underset{\sim}{G}}(\sigma)\cdot\sigma(c)\\ &= \sigma(a\cdot\chi_{\underset{\sim}{G}}(\sigma)\cdot c)\\ &= \sigma\varphi(c).\end{aligned}$$ □

<u>Remark</u>: Theorem 5.2 can directly be read off from [T] §4, corollary 2, at least if the residue field is perfect. But theorem 5.2 does not quite imply theorem 5.1. Here one would have to restrict both sides to an open subgroup.

References

[Be] Berthelot, P., Breen, L., Messing, W.: Theorie de Dieudonné Cristalline II, Springer LNM 930, (1982)

[Bo] Bourbaki, N.: Algèbre, Hermann (1961)

[Bon] Bourbaki, N.: Algêbre commutative, Hermann (1961)

[Ca] Cartier, P.: Groupes algebriques et groupes formels, Colloque CBRM, Brussels (1962), pp. 87-111

[C,F] Cassels, J.W.S., Fröhlich, A.: Algebraic number theory, Academic Press (1967)

[G] Gabriel, P.: Generalités sur les groupes algebriques Exposé IV_A in Seminaire de geometrie algebrique (1962/64) Springer LNM 151

[Ga] Gabriel, P.: Construction de preschemas quotient, Exposé V in Seminaire de geometrie algebrique (1962/64) Springer LNM 151

[G-D] Gabriel, P., Demazure, M.: Groupes algebriques, North Holland (1970)

[Gr] Grothendieck, A.: Éléments de geometrie algebrique IV, Publ. Mathematiques de l'IHES, No. 20

[H] Hartshorne, R.: Algebraic Geometry, Springer-Verlag (1977)

[Ha] Hazewinkel, M.: Formal groups and applications, Academic Press (1978)

[M] Milne, J.S.: Étale Cohomology, Princeton University Press (1980)

[Mu] Mumford, D.: Abelian Varieties, Oxford University Press (1970)

[M,F] Mumford, D., Fogarty, J.: Geometric invariant theory. Springer Verlag (1982)

[Oo] Oort, F.: Commutative group schemes, Springer Verlag LNM 15 (1966)

[O] Oort, F., Tate, J.: Group schemes of prime order, Ann. scient. Ec. Norm. Sup., 4e serie, t.3, (1970), pp. 1-21

[R] Raynaud, M.: Schémas en groupes de type (p,...,p), Bull. Soc. math. France, 102, (1974), pp. 241-280

[Ra] Raynaud, M.: Passage an quotient par une relation d'équivalence plate, Proceedings of a conference on local fields, [Driebergen, 1966], pp. 78-85 Springer Verlag

[S] Schappacher, N.: Tate's conjecture on the endomorphisms of abelian varieties. Contribution to this volume

[Se] Serre, J.P.: Proprieté's galoisienne des points d'ordre fini des courbes elliptiques, Inventiones Math. (1972), vol. 15, pp. 259-331

[Ser] Serre, J.P.: Répresentation lineaires des groupes finis, Hermann, Paris (1971)

[T] Tate, J.: p-divisible groups. Proceedings of a conference on local fields [Driebergen, 1966], pp. 158-183, Springer-Verlag

[W] Wüstholz, G.: The finiteness theorems of Faltings. Contribution to this volume.

IV

TATE'S CONJECTURE ON THE ENDOMORPHISMS OF ABELIAN VARIETIES

Norbert Schappacher

Contents:

Following Faltings and using older arguments due to Tate and Zarhin, we shall deduce, from the diophantine result [F2], II 4.3, Tate's conjectural description of the endomorphisms of abelian varieties over number fields, in terms of ℓ-adic representations.

§ 1 Statements

Let K be a number field (of finite degree over $\mathbb{Q}$), and let A be an abelian variety defined over K. Put $g = \dim A$. For a prime number ℓ, and $n \geq 1$, denote by $A[\ell^n]$ the kernel of multiplication by ℓ^n on A, and write, as usual,

$$T_\ell(A) = \varprojlim_n A[\ell^n](\bar{K}) \; ; \quad V_\ell(A) = T_\ell(A) \otimes_{\mathbb{Z}_\ell} \mathbb{Q}_\ell \quad ,$$

where $\bar{K}$ is a fixed algebraic closure of K.

T_ℓ and V_ℓ actually define covariant functors in an obvious way. The absolute Galois group $\pi = \mathrm{Gal}(\bar{K}/K)$ acts on $T_\ell(A)$, resp. $V_\ell(A)$, by $\mathbb{Z}_\ell$-linear, resp. $\mathbb{Q}_\ell$-linear, continuous transformations.

The object of this article is to prove the following theorem, known as Tate's conjecture on the endomorphisms $\mathrm{End}_K A$ of A defined over K.

1.1 Theorem. (i) *The action of* π *on* $V_\ell(A)$ *is semi-simple.*

(ii) *The natural map*

$$\mathrm{End}_K A \otimes_{\mathbb{Z}} \mathbb{Z}_\ell \longrightarrow \mathrm{End}_{\mathbb{Z}_\ell[\pi]}(T_\ell(A))$$

is an isomorphism.

Remark: The following facts can be found, e.g., in [Mu1]:

(i) Since K has characteristic 0, $T_\ell(A)$ is a free $\mathbb{Z}_\ell$-module of rank $2g$.

(ii) If B is another abelian variety over K, the homomorphisms $\mathrm{Hom}_K(A,B)$ always form a free $\mathbb{Z}$-module of finite type, and the functor T_ℓ induces an *injection*

$$\mathrm{Hom}_K(A,B) \otimes_{\mathbb{Z}} \mathbb{Z}_\ell \hookrightarrow \mathrm{Hom}_{\mathbb{Z}_\ell}(T_\ell(A),T_\ell(B))$$

whose image has to be in the submodule

$$\mathrm{Hom}_{\mathbb{Z}_\ell}(T_\ell(A),T_\ell(B))^\pi = \mathrm{Hom}_{\mathbb{Z}_\ell[\pi]}(T_\ell(A),T_\ell(B))$$

fixed by π, because $u(x)^g = u(x^g)$, for all $g \in \pi$, $x \in A[\ell^\infty]$, if $u \in \mathrm{End}\, A$ is defined over K. So, the essential claim of 1.1(ii) is *surjectivity*.

<u>1.2 Corollary.</u> *For* A,B *as above, the natural map*

$$\mathrm{Hom}_K(A,B) \otimes_{\mathbb{Z}} \mathbb{Z}_\ell \longrightarrow \mathrm{Hom}_{\mathbb{Z}_\ell[\pi]}(T_\ell(A),T_\ell(B))$$

is an isomorphism.

<u>Proof:</u> Apply 1.1 to the abelian variety $A \times B$. - See [T1], lemma 3.

The following corollary used to be known as the *isogeny conjecture* for abelian varieties over K.

1.3 Corollary. *The following statements are equivalent.*

(i) A *and* B *are isogenous over* K .

(ii) $V_\ell(A) \cong V_\ell(B)$, *as* π*-modules.*

(iii) *For almost all primes* v *of* K , $L_v(A,s) = L_v(B,s)$.

(iv) *For all* v , $L_v(A,s) = L_v(B,s)$.

(v) *For almost all* v, $\mathrm{tr}(F_v | V_\ell(A)^{I_v}) = \mathrm{tr}(F_v | V_\ell(B)^{I_v})$.

(vi) *For all* v, $\mathrm{tr}(F_v | V_\ell(A)^{I_v}) = \mathrm{tr}\,(F_v | V_\ell(B)^{I_v})$.

Here, $L_v(A,s)$ is the Euler factor at v of the Hasse-Weil L-function of A over K :

$$L(A/K,s) = \prod_v L_v(A,s) \quad (\text{for } \mathrm{Re}(s) > \tfrac{3}{2}).$$

Let $I_v \subset \pi$ be an inertia subgroup at v , and $F_v \in \pi/I_v$ a Frobenius element at v. Then the action of F_v on $T_\ell(A)^{I_v}$ is well-defined, and we put

$$L_v(A,s) = \frac{1}{\det(1 - \mathbb{N}v^{-s} \cdot F_v \mid T_\ell(A)^{I_v})} ,$$

$\mathbb{N}v$ being the cardinality of the residue class field at v . - This definition of L_v does not depend on the choice of the prime number $\ell \nmid \mathbb{N}v$, and I_v acts trivially on $T_\ell(A)$ for almost all v . Cf. [ST].

Corollary 1.3 asserts in particular that *the* L-*function* L(A/K,s) *is a complete isogeny invariant of* A/K .

<u>Proof of 1.3:</u> (i) $\Longleftrightarrow$ (ii). $f \in Hom(A,B)$ is an isogeny if and only if $T_\ell(f)$ has full rank, i.e., det $T_\ell(f) \neq 0$. This already implies (i) $\Longrightarrow$ (ii). On the other hand, suppose $\varphi: V_\ell(A) \to V_\ell(B)$ is an isomorphism of π-modules. Choose n such that $\ell^n \cdot \varphi \in Hom(T_\ell(A), T_\ell(B))$. This homomorphism comes from $Hom_K(A,B) \otimes_{\mathbb{Z}} \mathbb{Z}_\ell$, and can therefore be approximated by elements of $Hom(A,B)$. Since $\det(\ell^n\varphi) \neq 0$, the same will be true for good approximations. This way one finds the required isogeny.

<u>Remark:</u> Note that, for an isogeny $f: A \to B$, $T_\ell(f)$ is an isomorphism $T_\ell(A) \to T_\ell(B)$ if and only if $\ell \nmid \deg(f)$.

(v) $\Longrightarrow$ (ii) : A semi-simple representation of a $\mathbb{Q}_\ell$- algebra in a finite-dimensional $\mathbb{Q}_\ell$-vector space is determined by its character; [Bou], § 12, n°1. In our case, the character is continuous and therefore determined by its values on a dense subset of π. By Čebotarev's theorem (cf. [Se], chap. I), such a subset is provided by the Frobenius elements of a set of places of density 1.

The rest of the proof of 1.3 is logic. Note in particular that any quantifier may be used with ℓ in (ii).

<u>1.4 Remark</u> Since all higher étale cohomology groups

$$H^n_{ét}(A \times_K \overline{K}, \mathbb{Q}_\ell)$$

of the abelian variety A are given by exterior powers of

$$H^1_{ét}(A \times_K \overline{K}, \mathbb{Q}_\ell) \cong \mathrm{Hom}_{\mathbb{Q}_\ell}(V_\ell(A), \mathbb{Q}_\ell)$$

the semi-simplicity asserted in 1.1 implies that:

For all $n \geq 0$, *the action of* π *on* $H^n_{ét}(A \times_K \overline{K}, \mathbb{Q}_\ell)$ *is semi-simple.*

In fact, since the representations of π in question are in finite dimensional vector spaces over a field of characteristic 0 , this follows by passing to Lie-algebras: see [Hum], 13.2; [BoL], chap. I, § 6 n°5; cf. [BoL], chap. III, §9 n°8.

1.5 Tate's general conjecture

Let k be a field which is of finite type over its prime field, $\overline{k}$ a fixed algebraic closure of k , $\pi = \mathrm{Aut}_k(\overline{k})$ and ℓ a prime number different from the characteristic of k . Let X be a smooth projective geometrically connected variety over k , and write $\overline{X} = X \times_k \overline{k}$. Every closed irreducible subvariety $\overline{Z}$ of $\overline{X}$ of codimension r defines an ℓ-adic cohomology class

$$cl(\overline{Z}) \in H^{2r}(\overline{X}, \mathbb{Q}_\ell)(r) = \{\varprojlim_n H^{2r}_{ét}(\overline{X}, (\mu_{\ell^n})^{\otimes r})\} \otimes_{\mathbb{Z}_\ell} \mathbb{Q}_\ell ,$$

namely the image of $1 \in \mathbb{Q}_\ell$ under the natural map from relative cohomology

$$\mathbb{Q}_\ell \cong H^{2r}_{\overline{Z}}(\overline{X},\mathbb{Q}_\ell)(r) \longrightarrow H^{2r}(\overline{X},\mathbb{Q}_\ell)(r) .$$

Cf. [Mil], chap. VI.

Call $\mathcal{Z}^r(X)$ the free abelian group on subvarieties Z of X of codimension r *defined over* k , and

$$\mathfrak{A}^r(X) = \mathcal{Z}^r(X)/\text{kernel } (Z \mapsto c\ell(\overline{Z})).$$

Then the general form of Tate's conjecture related to our theorem is:

<u>Conjecture:</u> $\mathfrak{A}^r(X) \otimes_{\mathbb{Z}} \mathbb{Q}_\ell \xrightarrow{\cong} H^{2r}(\overline{X}, \mathbb{Q}_\ell)(r)^{\pi}.$

Cf. [T3].

We shall now indicate how theorem 1.1(ii) can be seen to be a special case of this conjecture. In fact, things become more transparent when we deduce corollary 1.2 instead. So, suppose A and B are abelian varieties over k , and consider the diagram

$$\begin{array}{ccc}
\text{Hom}\,(A,B) & \xrightarrow{(1)} & \text{Pic}^\circ(A \times B^*) \\
 & & \downarrow (2) \\
\downarrow (6) & & H^2(A \times B^*,\mathbb{Q}_\ell)(1) \\
 & & \downarrow (3) \\
 & & H^1(A,\mathbb{Q}_\ell) \otimes_{\mathbb{Q}_\ell} H^1(B^*,\mathbb{Q}_\ell)(1) \\
 & & \downarrow (4) \\
\text{Hom}_{\mathbb{Q}_\ell}(V_\ell(A),V_\ell(B)) & \xleftarrow{(5)} & V_\ell(A)^* \otimes V_\ell(B)
\end{array}$$

where B^*/k is the dual of B, and the maps are given as follows.

(1) For $\varphi \in \mathrm{Hom}(A,B)$, pullback of the Poincaré bundle $B \times B^*$ via $\varphi \times \mathrm{id}: A \times B^* \to B \times B^*$.

(2) First Chern class.

(3) Projection onto the (1,1) - component in the Künneth-decomposition.

(4) Use that $H^1(A,\mathbb{Q}_\ell) = V_\ell(A)^*$ (dual), and that the Weil-pairing on $V_\ell(B)$ induces a duality

$$H^1(B,\mathbb{Q}_\ell) \times H^1(B^*,\mathbb{Q}_\ell) \longrightarrow \mathbb{Q}_\ell(-1) ,$$

and thus an isomorphism

$$H^1(B^*,\mathbb{Q}_\ell)\ (1) \cong H^1(B,\mathbb{Q}_\ell)^* = V_\ell(B) .$$

(5) $\lambda \otimes b \mapsto (a \mapsto \lambda(a).b)$.

(6) Our natural map, induced by the functor V_ℓ.

It is easy to see that this diagram commutes. All maps are π-equivariant, and from the definition of the Poincaré bundle, it is clear that the image of $\mathrm{Hom}_k(A,B)$ under (3)∘(2)∘(1) is precisely $\mathfrak{A}^{1\otimes 1}(A \times B^*) \subset [H^1(A)\otimes H^1(B)(1)]^\pi$, the $H^1\otimes H^1$-projection of $\mathfrak{A}^1(A\times B^*)$. So, assuming Tate's conjecture, the surjectivity of (6) follows from the fact that (4) and (5) are isomorphisms.

1.6 A glance at the history

Elliptic curves over finite fields have lots of endomorphisms. This phenomenon was systemtically perused by Deuring in [Deu], and, as Tate points out in [T1], Deuring's results allow one to deduce the analogue of Corollary 1.2 for A,B elliptic curves over a *finite* field K (of characteristic $\neq \ell$). In [T1], Tate generalized this to abelian varieties over finite fields. In this case, the semi-simplicity of the π-action can be shown directly, but the pattern of proof developed by Tate turned out to be adequate even for the number field case. In a sequence of papers - [Z1] through [Z5] - Zarhin proved the analogue of 1.1 for most function fields of finite transcendence degree over a finite field. For this, he had to refine Tate's way of reducing 1.1 to a diophantine statement, and some of our reduction steps are inspired by Zarhin's refinements.

There have been partial results in the number field case before Faltings' general proof of 1.1, of which we mention Serre's results on elliptic curves (see [Se]), the case of complex multiplication (see [Shim], cf. [ZZ]), and the Jacobian of modular curves ([Ri]).

§2 Reductions

In this section, theorem 1.1 will be seen to be a consequence of a diophantine result on abelian varieties over K . Using the finiteness theorem [F2], II 4.3, this diophantine statement is seen to result from the behaviour of the modular height under certain isogenies. These height calculations will be performed in § 3.

The notations are those of the beginning of § 1.

(2.1) *To prove* 1.1(ii), *it suffices to show that the natural injection*

$$\mathrm{End}_K A \otimes_{\mathbb{Z}} \mathbb{Q}_\ell \longrightarrow \mathrm{End}_{\mathbb{Q}_\ell[\pi]}(V_\ell(A))$$

is an isomorphism.

In fact, this map is still injective since $\mathbb{Q}_\ell$ is flat over $\mathbb{Z}_\ell$. Furthermore, the cokernel of the $\mathbb{Z}_\ell$-linear map is torsion-free: an endomorphism of A vanishing on $A[\ell]$ is divisble by ℓ.

(2.2) *Let* $K' \supset K$ *be a finite extension. If 1.1 is true for* $A \times_K K'$ *over* K', *then it holds also over* K .

Let $\pi' = \mathrm{Gal}(\overline{K}/K')$, $\pi'' = \mathrm{Gal}(\overline{K}/K'')$, where K'' is a finite Galois extension of K containing K' . Since π'' is normal in π', the semi-simplicity of $V_\ell(A \times_K K') = V_\ell(A)$ as a π'-module implies that of the π''-module $V_\ell(A)$. π acts on the decomposition of this π''-module into simple factors; and adding up these π-orbits decomposes $V_\ell(A)$ as a π-module. Any $\varphi \in \mathrm{End}(T_\ell(A))$ fixed by π is also fixed by π'; therefore comes from an $f \in \mathrm{End}_{K'}(A \times_K K') \otimes_{\mathbb{Z}} \mathbb{Z}_\ell$. But f is again fixed under π, and thus lies in $\mathrm{End}_K A \otimes_{\mathbb{Z}} \mathbb{Z}_\ell$.

(2.3) *In proving 1.1, we may assume that* A *has semi-stable reduction over the ring of integers* $\mathcal{O}$ *of* K .

This is a consequence of 2.2 and Grothendieck's semi-stable reduction theorem - [Groth], thm. 3.6 - which asserts that there is a finite (separable) extension K' of K such that $A \times_K K'$ acquires semi-stable reduction over $\mathcal{O}_{K'}$. We shall recall the definition and various properties of abelian varieties with semi-stable reduction in § 3.

(2.4) *To prove 1.1, it suffices to show the following:*

$$(*)\begin{cases} \text{For every } \pi\text{-invariant subspace } W \subset V_\ell(A)\text{, there is} \\ u \in \mathrm{End}_K A \otimes_{\mathbb{Z}} \mathbb{Q}_\ell \text{ such that } u.\, V_\ell(A) = W. \end{cases}$$

A reduction step of this kind is already essential in Tate [T1]. Cf. also [Z4], lemma 3.1. First note that the right ideal

$$\{v \in \mathrm{End}_K A \otimes_{\mathbb{Z}} \mathbb{Q}_\ell \mid v.\, V_\ell(A) \subset W\} ,$$

like any right ideal in a semi-simple algebra, is generated

by some projector $u_\circ$, i.e., $u_\circ^2 = u_\circ$. If u exists as in (*), it follows that $u_\circ \cdot V_\ell(A) = W$. So every π-invariant subspace of $V_\ell(A)$ is a direct factor, which implies the semi-simplicity of the π-action.

Let C be the commutant of $\mathrm{End}_K A \otimes \mathbb{Q}_\ell$ in $\mathrm{End}_{\mathbb{Q}_\ell}(V_\ell(A))$. The commutant $C°$ of C equals $\mathrm{End}_K A \otimes \mathbb{Q}_\ell$, by the theorem of bicommutation - [Bou], § 5,n°4 -, again because $\mathrm{End}\, A \otimes \mathbb{Q}_\ell$ is a semi-simple algebra.

Assume we know (*) for all abelian varieties over K , in particular for $A \times A$. Then the graph

$$W = \{(x,\varphi(x)) \mid x \in V_\ell(A)\} \subset V_\ell(A)^2 = V_\ell(A \times A)$$

of any $\varphi \in \mathrm{End}_{\mathbb{Q}_\ell[\pi]}(V_\ell(A))$ is a π-invariant subspace, so there is $u \in \mathrm{End}_K A^2 \otimes \mathbb{Q}_\ell$ such that $u.\, V_\ell(A \times A) = W$.
It will be enough to show that $\varphi \in C°$. So take $\alpha \in C$. Then $\begin{pmatrix} \alpha & 0 \\ 0 & \alpha \end{pmatrix} \in \mathrm{End}(V_\ell(A)^2)$ commutes with $\mathrm{End}_K A^2 \otimes \mathbb{Q}_\ell$, in particular with u. Consequently $\begin{pmatrix} \alpha & 0 \\ 0 & \alpha \end{pmatrix} W \subset W$, which means that $\alpha\varphi = \varphi\alpha$, i.e., $\varphi \in C°$.

2.5 *Subspaces and ℓ-divisible groups.*

Given a $\mathbb{Q}_\ell$-linear subspace $W \subset V_\ell(A)$, put $U = W \cap T_\ell(A)$. Then, for $n \geq 1$,

$$\ell^{-n}\, U/U \hookrightarrow \ell^{-n}\, T_\ell(A)/T_\ell(A) = A\,[\ell^n](\bar{K})$$

defines the levels of an ℓ-divisible subgroup G of $A(\ell)/\bar{K}$ with height(G) = $\dim_{\mathbb{Q}_\ell} W$. (Cf. [Grun].) If W is π-invariant, G is defined over K.

Over K, we can divide A by G_n (for $n \geq 1$), obtaining abelian varieties A/G_n over K, together with isogenies

$$A \underset{f_n}{\overset{p_n}{\rightleftarrows}} A/G_n$$

of degree $\ell^{n.\dim W}$, such that

$$\begin{aligned} T_\ell(p_n)^{-1} \ (T_\ell(A/G_n)) &= \ell^{-n} U + T_\ell(A) \ , \\ T_\ell(f_n) \ (T_\ell(A/G_n)) &= U + \ell^n T_\ell(A) =: T_n \ . \end{aligned}$$

<u>(2.6)</u> *Given a π-invariant subspace $W \subset V_\ell(A)$, condition (*) of (2.4) is satisfied, if infinitely many of the abelian varieties $A/G_n (n \geq 0)$ are isomorphic to each other over K.*

The proof of 2.6 is the essential step which enabled Tate to prove the analogue of 1.1 for abelian varieties over finite fields; see [T1], Proposition 1.

To prove 2.6, let I be an infinite subset of $\mathbb{N}$, with smallest element $i_\circ$, such that, for all $i \in I$, there are isomorphisms defined over K,

$$v_i : A/G_{i_\circ} \xrightarrow{\sim} A/G_i \ .$$

In $\mathrm{End}_K A \otimes \mathbb{Q}_\ell$, consider the element u_i composed of

$$A \xrightarrow{f_{i_o}^{-1}} A/G_{i_o} \xrightarrow{v_i} A/G_i \xrightarrow{f_i} A \quad .$$

Viewed in End $V_\ell(A)$, u_i maps T_{i_o} onto $T_i \subset T_{i_o}$, in the notations of 2.5. But End T_{i_o} is compact. So, selecting a smaller I if necessary, we may assume that the sequence $(u_i)_{i \in I}$ converges to a limit u which still comes from $\mathrm{End}_K\, A \otimes \mathbb{Q}_\ell$ since this set is closed in End $V_\ell(A)$. Consider $U = \bigcap_{i \in I} T_i$. Since $u_i(T_{i_o}) = T_i$, every $x \in U$ is a limit $\lim_{i \in I} u_i(y_i)$, for certain $y_i \in T_{i_o}$. Passing to an accumulation point y of the y_i's we see that $U = u(T_{i_o})$. Thus, $u.\, V_\ell(A) = W$, as required.

Taking into account (2.3), it is now obvious that we will be done with the proof of Theorem 1.1, once we have obtained the following two results.

2.7 Proposition: *In the notation of* (2.5), *assuming* A, *and therefore all the* A/G_n, *to have semi-stable reduction, the modular height* $h(A/G_n)$ *is independent of* n, *for* n *sufficiently large.*

2.8. Theorem: *Given* g *and* c, *there exist, up to isomorphism, only finitely many abelian varieties* A *with semi-stable reduction over* K *such that* $\dim A = g$ *and* $h(A) \leq c$.

The proof of (2.7) and the reduction of (2.8) to the analogous statement for principally polarized abelian varieties which was proved in [F2] will be the subject of the next section.

§ 3 Heights

Before turning to the proofs proper of (2.7) and (2.8), let us recall some basic facts about abelian varieties with semi-stable reduction. The reference for this is [Groth].

Given an abelian variety A_K over the number field K, recall that there exists the *Néron-model* $\mathcal{A}$ of A_K which is a smooth group scheme over the ring of integers R of K, and is uniquely characterized by the fact that

$$\mathrm{Hom}_R(S,\mathcal{A}) \cong \mathrm{Hom}_K(S_K,A_K) ,$$

for every smooth group scheme S over R with generic fibre S_K. *From now on, we will always denote by* A *the connected component of* $\mathcal{A}$, with fibres the connected components of 0 of the fibres of $\mathcal{A}$.

A_K is said to have *semi-stable reduction over* K, if for every $s \in \mathrm{Spec}\, R$, the fibre A_s sits in an exact sequence

$$1 \longrightarrow T_s \longrightarrow A_s \longrightarrow B_s \longrightarrow 0 ,$$

with an abelian variety B_s and a torus T_s over $k(s)$. Equivalently, [Groth], 3.2, A_K has semi-stable reduction, if there exists some smooth separated group scheme G of finite type over $\mathrm{Spec}\, R$ whose fibres are all extensions of an abelian variety by a torus as above, and whose generic fibre is A_K.

Assume now that A_K and B_K are abelian varieties with semi-stable reduction over K. Suppose an isogeny

$$\varphi : A_K \longrightarrow B_K$$

over K is given. By the universal property of the (connected) Néron model, φ certainly extends to a morphism over Spec R:

$$\varphi : A \longrightarrow B \; .$$

Semi-stability implies furthermore that this morphism is *faithfully flat*, and that the kernel

$$G = \ker (A \xrightarrow{\varphi} B)$$

is a quasi-finite, flat group scheme over Spec R. (Cf.[Groth], 2.2.1, or [Mu2], lemma 6.12 : the typical bad case ruled out by semi-stability is multiplication by $p : \mathbb{G}_a \longrightarrow \mathbb{G}_a$, over a field of characteristic p.) Note that G *is not necessarily a finite group scheme over* Spec R (unless A and B have good reduction everywhere) : its fibres will have varying orders in general.

At any rate, one obtains the exact sequence

$$0 \longrightarrow s^*(\Omega^1_{B/R}) \xrightarrow{\varphi^*} s^*(\Omega^1_{A/R}) \longrightarrow s^*(\Omega^1_{G/R}) \longrightarrow 0.$$

Here, s denotes the zero-sections of the group schemes in question. The exactness at the centre follows from that of the well-known sequence of relative differentials,

$$\varphi^*(\Omega^1_{B/R}) \longrightarrow \Omega^1_{A/R} \longrightarrow \Omega^1_{A/B} \longrightarrow 0 .$$

Now, the order of the finite group $s^*(\Omega^1_{G/R})$ equals

$$\#(s^*\Omega^1_{G/R}) = \#\operatorname{coker}(\Lambda^g \varphi^* : \omega_{B/R} \longrightarrow \omega_{A/R}),$$

where $\omega_{X/R}$ denotes the maximal exterior power of $s^*(\Omega^1_{X/R})$. This is shown by localizing and applying a well-known corollary of the theorem of elementary divisors.

Recall the definition of the *modular height of a (semi-)abelian variety*:

$$h(A) = \frac{1}{[K:\mathbb{Q}]} \deg(\omega_{A/R}) ,$$

with:

$$\deg(\omega_{A/R}) = \log \#(\omega_{A/R}/p.R) - \sum_{v|\infty} \varepsilon_v . \log \|p\|_v ,$$

p being a non-zero element of $\omega_{A/R}$, and $\varepsilon_v = 1$ or 2, according as v is real or complex.

As φ changes the volume by $\sqrt{\deg \varphi}$ at every infinite place of K, we see that we have the

<u>(3.1) Isogeny Formula:</u> *Under the above assumptions,*

$$h(B) - h(A) = \frac{1}{2}\log(\deg\varphi) - \frac{1}{[K:\mathbb{Q}]}\log\#(s^*\Omega^1_{G/R}) .$$

<u>(3.2)</u> For the application of this isogeny formula in the proof of (2.7) we shall need the theory of the *fixed and torus parts* of $T_\ell(A_K)$, for an abelian variety A_K with semi-stable

reduction. See [Groth], esp. § 5. Let us recall the basics of this theory in the situation we shall encounter.

Let v be a place of K dividing ℓ, and R_v the completion of R at v. As over the spectrum of any Henselian local ring, every quasi-finite scheme X over Spec R_v decomposes as

$$X = \tilde{X} \amalg Y ,$$

where $\tilde{X}$ is *finite* over R_v, and Y has no special fibre, cf. [EGA II]6.2.6. Given A_K with semi-stable reduction as before, we can apply this to the quasi-finite group scheme $A[\ell^\nu]$, the kernel of multiplication by ℓ^ν on the connected Nêron model of A_K, considered over the completion R_v, thus obtaining its *finite part* $\widetilde{A[\ell^\nu]}$ over R_v. These finite parts make up a strict (i.e., $\ell: A \to A$ is surjective) projective system which then defines what is called the *fixed part* of the Tate-module of A:

$$T_\ell(A)^f \subset T_\ell(A) .$$

We shall make use of this submodule *in the generic fibre* (i.e., the only Tate-module we ever considered in §§ 1 and 2) which may be written all explicitely

$$T_\ell(A_K)^f(\overline{K_v}) \subset T_\ell(A_K)(\overline{K_v}) .$$

Henceforth, we shall simply write

$$T_\ell(A_{K_v})^f \subset T_\ell(A_{K_v}) ,$$

even if we think only of the ℓ-adic Galois-representation given by the $\overline{K_v}$-rational points.

Let $\hat{A}$ over $Spf(R_v)$ be the formal completion of A/R_v along its special fibre $A_\circ$. Now, in the decomposition above

$$A[\ell^\nu] = \widetilde{A[\ell^\nu]} \amalg C_\nu \qquad (\nu \geq 0)$$

we have

$$\hat{A}[\ell^\nu] = \widehat{\widetilde{A[\ell^\nu]}} ,$$

because C_ν has no special fibre. Therefore,

$$T_\ell(\hat{A}) = T_\ell(A)^f ,$$

if we agree to identify finite schemes over $Spec\ R_v$ with finite formal schemes over $Spf\ (R_v)$. (Cf. [EGA III],4.8.)

Furthermore, by semi-stability, the special fibre $A_\circ$ sits in an exact sequence

$$1 \longrightarrow T_\circ \longrightarrow A_\circ \longrightarrow B_\circ \longrightarrow 0 ,$$

for some abelian variety $B_\circ$ and torus $T_\circ$ over $k_v = R_v/\mathfrak{m}_v$. For every $n \geq 1$, there is a unique torus T_n over $R_v/\mathfrak{m}_v^{(n+1)}$ with special fibre $T_\circ$ ([Gro], 3.6 bis). Being unique, the T_n fit together to define a formal torus $\hat{T}/R_v$ which injects into $\hat{A}$. This torus gives us a submodule

$$T_\ell(A)^t := T_\ell(\hat{T}) \subset T_\ell(\hat{A}) = T_\ell(A)^f .$$

Here too, we can consider the generic fibre. So we have a two-step filtration

$$T_\ell(A_{K_v})^t \subset T_\ell(A_{K_v})^f \subset T_\ell(A_{K_v})$$

of the Tate-module of the semi-stable abelian variety A_K over K .

Likewise, for the dual abelian variety A_K^* over K , we get submodules

$$T_\ell(A_{K_v}^*)^t \subset T_\ell(A_{K_v}^*)^f \subset T_\ell(A_{K_v}^*) \quad .$$

The Weil pairing provides an alternating duality

$$T_\ell(A_K) \times T_\ell(A_K^*) \longrightarrow \mathbb{Z}_\ell(1) \quad .$$

The *Orthogonality Theorem* - [Groth], 5.2 - asserts that, with respect to this paring,

$$T_\ell(A_{K_v})^t = (T_\ell(A_{K_v}^*)^f)^\perp \quad ,$$

and, of course, the other way around:

$$T_\ell(A_{K_v}^*)^t = (T_\ell(A_{K_v})^f)^\perp \quad .$$

As a first consequence of this, let us note right away the

3.3 Lemma: *Call* $D_v = \mathrm{Gal}(\overline{K_v}/K_v) \subset \pi$ *the decomposition group and* $I_v \subset D_v$ *the inertia subgroup of* v . *Then* I_v *acts trivially on* $T_\ell(A_{K_v})/T_\ell(A_{K_v})^f$ *, and* D_v *acts via a finite quotient.*

Proof: By the orthogonality theorem,

$$T_\ell(A_{K_v})/T_\ell(A_{K_v})^f \cong \mathrm{Hom}(T_\ell(\hat{T}), T_\ell(\mathbb{G}_m)) \quad .$$

So, the lemma follows from the fact that $\hat{T}$ is split by a finite unramified extension of K_v (in fact, $T_\circ$ is split by the algebraic closure of the residue field k_v) .

(3.4) We can now return to the situation envisaged in (2.5), with a view to proving (2.7). Rewriting (2.5) in our present notation, we are given an abelian variety A_K with semi-stable reduction over K , an ℓ-divisible group $(G_{nK})_{n\geq 0}$, and the quotients

$$A_K \xrightarrow{p_n} (A_K/G_{nK}) = A_{nK} \quad .$$

Passing to connected Néron models, call G_n now the kernel of the isogeny of connected Néron models

$$p_n\colon A \longrightarrow A_n \qquad \text{over} \quad R \quad .$$

Fixing a place $v|\ell$, decompose, as in (3.2) above,

$$G_n = \tilde{G}_n \amalg H_n \qquad \text{over} \quad R_v \quad .$$

with $\tilde{G}_n$ finite over Spec R_v, and H_n without special fibre. - Thus,

$$\tilde{G}_n = \hat{A}[\ell^n] \cap G_n \quad .$$

Now, our problem is that $\bigcup_{n\geq 0} \tilde{G}_n$ *need not be an* ℓ-*divisible group over* R_v.

In fact, consider first the Galois representation in the generic fibre : $\bigcup_{n\geq 0} \tilde{G}_n(\overline{K_v})$. Being an intersection of two ℓ-divisible groups over K_v, this is of the form:

$$\begin{pmatrix} \overline{K_v}\text{-rational points of an} \\ \ell\text{-divisible group over } K_v \end{pmatrix} \oplus \begin{pmatrix} \text{finite abelian} \\ \text{group} \end{pmatrix} .$$

The finite group is contained in some $\tilde{G}_{n_0}(\overline{K_v})$, so for $\Gamma_n = \tilde{G}_{n_0+n}/\tilde{G}_{n_0}$ $(n\geq 0)$, we find that $\bigcup_{n\geq 0} \Gamma_n(\overline{K_v})$ is ℓ-divisible *over* K_v.

But $\bigcup_{n\geq 0} \Gamma_n$ need not be an ℓ-divisible group *over* R_v.
In fact, the sequences

$$0 \longrightarrow \Gamma_n \longrightarrow \Gamma_{n+m} \xrightarrow{\ell^n} \Gamma_m \longrightarrow 0$$

may not be exact *over* R_v. This problem is discussed on the last page of [T2] , and we are going to apply Tate's trick to get around it: Look at the maps induced by multiplication by ℓ

$$(*)_n : \Gamma_{n+2}/\Gamma_{n+1} \xrightarrow{\ell} \Gamma_{n+1}/\Gamma_n \qquad (n\geq 0) \quad .$$

Let E_n be the affine algebra of Γ_{n+1}/Γ_n . Since $\bigcup_{n\geq 0} \Gamma_n$ is an ℓ-divisible group *over* K_v, $F:=E_n \otimes_{R_v} K_v$ is a finite-dimensional K_v-algebra which does not depend on n . So, the E_n form an increasing sequence of orders in F. Such a sequence has to become stationary. In other words, the maps $(*)_n$ are isomorphisms for, say, $n \geq n_1$. We claim that the

$$\tilde{\Gamma} := \Gamma_{n_1+n}/\Gamma_{n_1} \cong \tilde{G}_{n_0+n_1+n}/\tilde{G}_{n_0+n_1} \qquad (n\geq 0)$$

constitute an ℓ-divisible group over R_v .- We have to show that the long rows of the following commutative diagram are exact, for all n .

$$\begin{array}{ccccccccc}
 & & 0 & \longrightarrow & \Gamma_{n_1+n+2}/\Gamma_{n_1+n+1} & \overset{(*)_{n_1+n}}{\cong} & \Gamma_{n_1+n+1}/\Gamma_{n_1+n} & & \\
 & & & & \uparrow & & \uparrow & & \\
0 & \longrightarrow & \tilde{\Gamma}_1 & \longrightarrow & \tilde{\Gamma}_{n+2} & \xrightarrow{\ \ell\ } & \tilde{\Gamma}_{n+1} & \longrightarrow & 0 \\
 & & \| & & \cup & & \cup & & \\
0 & \longrightarrow & \tilde{\Gamma}_1 & \longrightarrow & \tilde{\Gamma}_{n+1} & \xrightarrow{\ \ell\ } & \tilde{\Gamma}_n & \longrightarrow & 0
\end{array}$$

This follows from this very diagram by induction.

<u>(3.5)</u> We can now begin to *show that*

$$h(A_{n_0+n_1}) = h(A_{n_0+n_1+n})$$

for all $n\geq 0$, which gives (2.7) .

To simplify notations, let us pretend that $n_0=n_1=0$, so that

$\tilde{\Gamma}_n = \tilde{G}_n$. Recall that $A_n = A/G_n$ (connected semi-abelian scheme over R). From 3.1, we get:

$$h(A_n) - h(A) = \frac{1}{2} \log (\deg p_n) - \frac{1}{[K:\mathbb{Q}]} \log \#(s^*\Omega^1_{G_n/R}).$$

Recall (3.2) that, for all places v of K dividing ℓ,

$$G_n = \tilde{G}_n \amalg H_n \qquad \text{over } R_v ,$$

where H_n is concentrated in the generic fibre, and $\tilde{G}_n$ is finite over R_v. Completing along the special fibre, one finds $\hat{G}_n = \hat{\tilde{G}}_n$, over R_v. - Taking differentials commutes with completion, so we get successively:

$$\#(s^*\Omega^1_{G_n/R}) = \prod_{v|\ell} \#(s^*\Omega^1_{G_n/R_v}) = \prod_{v|\ell} \#(s^*\Omega^1_{\hat{G}_n/R_v}) = \prod_{v|\ell} \#(s^*\Omega^1_{\tilde{G}_n/R_v}).$$

By [Grun], 3.4, we have

$$\#(s^*\Omega^1_{\tilde{G}_n/R_v}) = \#(R_v/\ell^n R_v)^{d_v} ,$$

where d_v is the *dimension* of the ℓ-divisible group $\bigcup_{n\geq 0} \tilde{G}_n$ over R_v (we have assumed for simplicity that this *is* ℓ-divisible).

Call $h = \dim_{\mathbb{Q}_\ell}(W) = \operatorname{rank}_{\mathbb{Z}_\ell}(U)$ (see 2.5) the *height* of the ℓ-divisible group $\bigcup_{n\geq 0} G_{nK}$ *over* K.

We find:

$$h(A_n) - h(A) = n.\log(\ell).\left\{\frac{h}{2} - \sum_{v|\ell}\frac{[K_v : \mathbb{Q}_\ell]}{[K : \mathbb{Q}]} d_v\right\}$$

We have to show that the expression in curly brackets is zero!

(3.6) Put $\tilde{\pi} = \mathrm{Gal}(\overline{\mathbb{Q}}/\mathbb{Q})$, and consider the induced Galois-representations (recall that $U = T_\ell(\bigcup_n G_{nK})$, see 2.5)

$$\tilde{U} = \mathrm{Ind}_\pi^{\tilde{\pi}} U \subset \mathrm{Ind}_\pi^{\tilde{\pi}} T_\ell(A_K) = T_\ell(B_{\mathbb{Q}}) ,$$

where $B_{\mathbb{Q}} = \mathrm{Res}_{K/\mathbb{Q}}(A_K)$ is the abelian variety over $\mathbb{Q}$ obtained from A_K by Weil-restriction from K to $\mathbb{Q}$. We are going to more or less evaluate the character

$$\det \tilde{U} : \tilde{\pi} \longrightarrow \mathbb{Z}_\ell^*$$

in two different ways!

First, it is well-known (cf,,e.g.,[Mar], 3.2, which is easily generalized to our situation) that

$$\det \tilde{U} = \varepsilon^h . (\det U \circ \mathrm{Ver}_\pi^{\tilde{\pi}}) ,$$

where $\varepsilon : \tilde{\pi} \longrightarrow \{\pm 1\}$ is the signature of the permutations induced by $\tilde{\pi}$ on the homogeneous space $\tilde{\pi}/\pi$, and $\mathrm{Ver}_\pi^{\tilde{\pi}}$ is the transfer map : $\tilde{\pi}^{ab} \longrightarrow \pi^{ab}$. To compute $\det U$ at a place v of K dividing ℓ, up to an unramified character of finite order, we may replace $\bigcup_n G_{nK_v}$ by $\bigcup_n \tilde{G}_{nK_v}$ - this follows from (3.3) since

$$T_\ell(\underset{n}{\cup}G_{nK_v})/T_\ell(\underset{n}{\cup}\tilde{G}_{nK_v}) \hookrightarrow T_\ell(A_{K_v})/T_\ell(A_{K_v})^f \; .$$

Now, by [Grun], 5.2 , we have

$$\Lambda^{\tilde{h}}\, T_\ell(\cup\tilde{G}_{nK_v}) \otimes_{\mathbb{Z}_\ell} C_v \cong C_v(d_v) \quad ,$$

where $\tilde{h}$ is the *height* of the ℓ-divisible group $\underset{n}{\cup}\tilde{G}_n$ over R_v , and $C_v(d_v)$ is the completion of $\overline{K_v}$ with Galois-action given by the restriction to $\mathrm{Gal}(C_v/K_v) \hookrightarrow \pi\subset\tilde{\pi}$ of the character $\chi_\ell{}^{d_v}$, with $\chi_\ell : \tilde{\pi} \longrightarrow \mathbb{Z}_\ell^*$ the cyclotomic character giving the action of $\tilde{\pi}$ on $T_\ell(\mathbb{G}_m)$. Composing with $\mathrm{Ver}_\pi^{\tilde{\pi}}$, and adding up the results for all $v|\ell$, we see that

$$(\det \mathcal{U} \circ \mathrm{Ver}_\pi^{\tilde{\pi}}) \cdot \chi_\ell^{-\sum\limits_{v|\ell}[K_v:\mathbb{Q}_\ell]d_v}$$

is unramified at ℓ .(The transfer map does not introduce any new ramification because it corresponds to the natural map of ideles $\mathbb{Q}_{\mathbb{A}}{}^* \longrightarrow K_{\mathbb{A}}{}^*$, via class field theory.) On the other hand, at each finite place w of K not dividing ℓ , the inertia I_w acts unipotently on $\cup$ since A_K has semi-stable reduction: [Groth], 3.8. As unipotent matrices have determinant 1, we conclude that the character

$$\varphi = \det\tilde{\mathcal{U}} \cdot \varepsilon^{-h} \cdot \chi_\ell^{-\sum\limits_{v|\ell}[K_v:\mathbb{Q}_\ell]d_v} \quad : \tilde{\pi} \longrightarrow \mathbb{Z}_\ell{}^*$$

is unramified at every rational prime.

But $\mathbb{Q}$ has no (abelian) extensions that are unramified at all finite places (use Minkowski or class field theory). So, by class field theory, φ has to be the *trivial character*.

Thus for any rational prime $p \neq \ell$ where $B_{\mathbb{Q}}$ has good reduction, if $F_p \in \tilde{\pi}^{ab}$ is a Frobenius element at p, then, on the one hand, we certainly have $\varphi(p) = 1$. On the other hand, by the part of the "Weil-conjectures" proved by Weil himself, the eigenvalues of F_p on $\tilde{U}$ are algebraic numbers purely of absolute value $p^{1/2}$, since $\tilde{U} \subset T_\ell(B_{\mathbb{Q}})$. So, $\det \tilde{U}(F_p)$ is an algebraic number purely of absolute value $p^{h[K:\mathbb{Q}]/2}$ (recall that $h = \mathrm{rank}_{\mathbb{Z}_\ell}(U)$!). As $\chi_\ell(F_p) = p \in \mathbb{Z}_\ell{}^*$ we conclude that

$$\frac{h[K:\mathbb{Q}]}{2} = \sum_{v|\ell} [K_v:\mathbb{Q}_\ell]\, d_v \quad .$$

This proves (3.5), *and therefore* (2.7).

We still have to deduce the diophantine result 2.8 from the corresponding assertion, proved in [F2], about *principally polarized* abelian varieties. We claim it will be enough to establish the following two results:

<u>3.7 Proposition:</u> *For any abelian variety* A_K *over* K *with semi-stable reduction, calling* $A_K{}^*$ *its dual abelian variety, we have*

$$h(A_K{}^*) = h(A_K) \quad .$$

<u>3.8 Lemma [Zarhin]:</u> *For any abelian variety* A_K *over* K, *calling* $A_K{}^*$ *its dual,* $A_K{}^4 \times A_K{}^{*4}$ *carries a principal polarization.*

In fact, given 3.7 and 3.8, we find

$$h(A_K^4 \times A_K^{*4}) = 8 \cdot h(A_K) \quad ,$$

and of course,

$$\dim (A_K^4 \times A_K{*}^4) = 8 \dim (A_K) \quad .$$

So, the number of K-isomorphism classes of $A_K^4 \times A_K{*}^4$ (even equipped with a principal polarization) is finite. But the ring $\mathcal{E} = \mathrm{End}_K(A_K{}^4 \times A_K{*}^4)$ is finitely generated over $\mathbb{Z}$, and $\mathcal{E} \otimes \mathbb{Q}$ is a semi-simple algebra. Therefore there are, up to conjugation by $\mathcal{E}^*$, only finitely many idempotents in $\mathcal{E}$. (In fact: e and e' are conjugate if and only if $\mathcal{E}e \cong \mathcal{E}e'$ and $\mathcal{E}(1-e) \cong \mathcal{E}(1-e')$. But the number of subspaces $(\mathcal{E} \otimes \mathbb{Q}) \cdot e$ and $(\mathcal{E} \otimes \mathbb{Q})(1-e)$ is finite, and the theorem of Jordan and Zassenhaus implies there are only finitely many choices of a lattice in each of these spaces.) Thus, 2.8 follows from 3.7 and 3.8.

<u>Proof of 3.7 :</u> In computing h, we are free to make finite extensions of the base field. Also, the proposition is trivial if A_K is principally polarizable, because then $A \cong A^*$. Now, over a suitable extension field, A is isogenous to a principally polarized abelian variety. So, it is enough to show that $h(A^*) - h(A)$ is an isogeny invariant. Since every isogeny can be factored (over an extension field) into steps of prime degree, we are reduced to showing that

$$h(A^*) - h(B^*) + h(B) - h(A) = 0 \quad ,$$

provided there is an isogeny $\varphi: A \longrightarrow B$ of degree ℓ.
By our isogeny formula 3.1, applied to φ and to the dual isogeny

$$\varphi^* : B^* \longrightarrow A^* \quad \text{(also of degree } \ell\text{)} ,$$

with respective kernels $G \hookrightarrow A$ and $G^* \hookrightarrow B^*$, we have to prove that

$$[K:\mathbb{Q}] \,.\, \log(\ell) = \log\left(\#(s^*\Omega^1_{G/R}) \,.\, \#(s^*\Omega^1_{G^*/R})\right) .$$

Using the localisation and completion process as in (3.5), it suffices to show that, for every place v of K dividing ℓ,

$$\underline{(3.9)} \quad \#(s^*\Omega^1_{\hat{G}/R_v})\, \#(s^*\Omega^1_{\hat{G^*}/R_v}) = \#(R_v/\ell R_v) .$$

To prove 3.9, we shall break up φ and φ^* according to the two-step filtrations of T_ℓ discussed in 3.2. - $T_\ell(\varphi)$ and its dual $T_\ell(\varphi^*)$ induce three pairs of dual maps (the duality following from the orthogonality theorem quoted in 3.2) :

$$\text{(I)} \quad \begin{array}{ccc} T_\ell(A)^t & \longrightarrow & T_\ell(B)^t \\ T_\ell(A^*)/T_\ell(A^*)^f & \longleftarrow & T_\ell(B^*)/T_\ell(B^*)^f \end{array}$$

$$\text{(II)} \quad \begin{array}{ccc} T_\ell(A)^f/T_\ell(A)^t & \longrightarrow & T_\ell(B)^f/T_\ell(B)^t \\ T_\ell(A^*)^f/T_\ell(A^*)^t & \longleftarrow & T_\ell(B^*)^f/T_\ell(B^*)^t \end{array}$$

$$\text{(III)}\quad \begin{array}{ccc} T_\ell(A)/T_\ell(A)^f & \longrightarrow & T_\ell(B)/T_\ell(B)^f \\ T_\ell(A^*)^t & \longleftarrow & T_\ell(B^*)^t \end{array}$$

Considering the decompositions of the formal completions of our semi-stable abelian varieties over R_v:

$$\begin{array}{ccccccccc} 1 & \longrightarrow & \hat{T}(A) & \longrightarrow & \hat{A} & \longrightarrow & \widehat{Ab}(A) & \longrightarrow & 0 \\ & & \downarrow{\scriptstyle \hat{T}(\varphi)} & & \downarrow{\scriptstyle \hat{\varphi}} & & \downarrow{\scriptstyle \widehat{Ab}(\varphi)} & & \\ 1 & \longrightarrow & \hat{T}(B) & \longrightarrow & \hat{B} & \longrightarrow & \widehat{Ab}(B) & \longrightarrow & 0 \end{array},$$

the maps between the torus parts of the Tate-modules in (I) and (III) are induced by the map $\hat{T}(\varphi)$ between the completed tori (resp. by $\hat{T}(\varphi^*)$), and the maps in (II) are derived from the pair of dual mappings $\widehat{Ab}(\varphi)$, $\widehat{Ab}(\varphi^*)$ between formal abelian schemes over $\mathrm{Spf}(R_v)$.

$\hat{G}$ and $\hat{G^*}$ have order 1 or ℓ, so *precisely one* of the three pairs of dual maps will have non-trivial kernels. More precisely: Suppose a kernel sits in (I). Then $\hat{G} \subset \hat{T}(A)$, and forcibly $\hat{G^*} = 0$. As $\hat{G}$ is of multiplicative type,

$$\#(s^*\Omega^1_{\hat{G}/R_v}) = \#\,(R_v/\ell R_v)$$

- just as for μ_ℓ, see [Grun], 2.5. Next, suppose $\hat{G} \not\subset \hat{T}(A)$, and $\hat{G^*} \neq 0$. Then $\hat{T}(\varphi)$ and $\hat{T}(\varphi^*)$ are isomorphisms, whereas $\widehat{Ab}(\varphi)$ and $\widehat{Ab}(\varphi^*)$ are dual isogenies of degree ℓ , with kernels $\hat{G}$ and $\hat{G}^*$, respectively. Applying the functor $\mathrm{Hom}(.,\hat{\mathbb{G}}_m)$ to the short exact sequence

$$0 \longrightarrow \hat{G} \longrightarrow \hat{Ab}(A) \longrightarrow \hat{Ab}(B) \longrightarrow 0 \quad ,$$

we obtain the exact sequence (of fppf-sheaves)

$$\begin{array}{ccccc} 0 \longrightarrow \mathrm{Hom}(\hat{G},\mathbb{G}_m) & \longrightarrow & \mathrm{Ext}^1(\hat{Ab}(B),\hat{\mathbb{G}}_m) & \longrightarrow & \mathrm{Ext}^1(\hat{Ab}(A),\hat{\mathbb{G}}_m) \\ & & \| & & \| \\ & & \hat{Ab}(B^*) & & \hat{Ab}(A^*) \end{array} .$$

This shows that $\hat{G}$ and $\hat{G}^*$ are dual to each other, and consequently (see [Grun], 2.4) :

$$\#\,(s^*\Omega^1_{\hat{G}/R_v})\ \#(s^*\Omega^1_{\hat{G}^*/R_v}) = \#(R_v/\ell R_v) \quad ,$$

as required.

Finally, if the maps in (I) and (II) are all bijective, then we must have $\hat{G} = 0$ and $\hat{G}^* \subset \hat{T}(B^*)$. This case is exactly dual to the first one we treated.

q.e.d.

To complete this section, we still have to do the

<u>Proof of lemma 3.8:</u>

There is always some polarization on A_K over K , so let $\mathcal{L}$ be an ample line bundle on A_K defined over K, giving rise to the symplectic form

$$<,> : T_\ell(A_K) \times T_\ell(A_K) \longrightarrow \mathbb{Z}_\ell(1) \quad ,$$

for any prime ℓ. Choose an integer $N > 0$ such that, for all ℓ,

$$T_\ell(A_K)^* \subset \frac{1}{N} T_\ell(A_K) \subset T_\ell(A_K) \otimes_{\mathbb{Z}_\ell} \mathbb{Q}_\ell \quad ,$$

where $T_\ell(A_K)^*$ is the dual lattice of $T_\ell(A_K)$ with respect to $<,>$. (E.g., $N = \deg(\mathcal{L})$.) There are $a,b,c,d \in \mathbb{Z}$ with

$$a^2 + b^2 + c^2 + d^2 \equiv -1 \pmod N.$$

(In fact, $2^2 + 1^2 + 1^2 + 1^2 \equiv -1 \pmod 8$, and if $-1 \notin (\mathbb{F}_p^*)$, then $1 \notin -(\mathbb{F}_p^*)^2 \cup 1 + (\mathbb{F}_p^*)^2$, so that $-(\mathbb{F}_p^*)^2 \cap 1 + (\mathbb{F}_p^*)^2 \neq \emptyset$. From there, one goes with Newton.) Put

$$\alpha = \begin{pmatrix} a & -b & -c & -d \\ b & a & d & -c \\ c & -d & a & b \\ d & c & -b & a \end{pmatrix} \in M_4(\mathbb{Z}) \quad ,$$

so that ${}^t\alpha \cdot \alpha \equiv -1 \pmod N$. For each ℓ, consider the lattice

$$\begin{pmatrix} I_4 & \alpha \\ 0 & I_4 \end{pmatrix} (T_\ell(A_K)^4 \oplus T_\ell(A_K)^{*4}) \subset V_\ell(A_K)^8 \quad .$$

It is easily checked that, by its very construction, this lattice is selfdual and integral-valued with respect to the form $\langle\, ,\rangle^8$ on $V_\ell(A_K)^8$. (Note that, as α has rational-integral entries, the Rosati involution of $\langle,\rangle^4$ on α is simply the transpose.) As the lattice is clearly Galois-invariant, there is a quotient B_K of A_K over K, such that $T_\ell(B_K)$ is the above lattice. B_K is obviously isomorphic to $A_K^4 \times A_K^{*4}$, and from the properties of $T_\ell(B_K)$ we see that it admits a principal polarization.

q.e.d.

This completes the proof of the Tate conjecture.

§ 4 Variants

In this section, we collect some variants of Theorem 1.1 , and indicate a possible variation of its proof.

Let us start with the following obvious consequence of Theorem 1.1 and Corollary 1.2. The notations are those of the beginning of § 1.

4.1 Variant *Let* T *be a finite set of rational primes. Then:*

(i) *The action of* π *on* $\bigoplus_{\ell \in T} V_\ell(A)$ *is semi-simple.*

(ii) *The natural map*

$$\mathrm{Hom}_K(A,B) \otimes_{\mathbb{Z}} \Big(\prod_{\ell\in T} \mathbb{Z}_\ell\Big) \longrightarrow \prod_{\ell\in T} \mathrm{Hom}_\pi(T_\ell(A), T_\ell(B))$$

is an isomorphism.

There is a less trivial and more interesting way to pass from one $\mathbb{Z}_\ell$ to $\hat{\mathbb{Z}} = \varprojlim_{n\in\mathbb{N}} (\mathbb{Z}/n\,\mathbb{Z}) = \prod_{\text{all }\ell} \mathbb{Z}_\ell$:

4.2 Theorem (See last remark of [F1];cf .[De],2.7) *Let*

$$T(A) = \prod_{\text{all}\,\ell} T_\ell(A) \;, \quad \text{and}$$

$$\rho : \hat{\mathbb{Z}}[\pi] \longrightarrow \mathrm{End}_{\hat{\mathbb{Z}}}(T(A))$$

be the homomorphism given by the action of π *on* T(A) . *Then the subalgebra* $\rho(\hat{\mathbb{Z}}[\pi])$ *of* $\mathrm{End}_{\hat{\mathbb{Z}}}(T(A))$ *is of finite index in the commutant of*

$$\mathrm{End}_K(A) \hookrightarrow \mathrm{End}_{\hat{\mathbb{Z}}}(T(A))$$

in $\mathrm{End}_{\hat{\mathbb{Z}}}(T(A))$.

Note that 4.2 implies 1.1. In fact, 4.2 implies that, for all primes ℓ , the image of

$$\rho_\ell \otimes \mathbb{Q}_\ell : \mathbb{Q}_\ell[\pi] \longrightarrow \mathrm{End}_{\mathbb{Q}_\ell}(V_\ell(A))$$

is the commutant of the semi-simple $\mathbb{Q}_\ell$-algebra $\mathrm{End}_K A \otimes_{\mathbb{Z}} \mathbb{Q}_\ell$. So, this image is itself a semi-simple $\mathbb{Q}_\ell$-algebra, whence (i) of 1.1. Furthermore, by the theorem of bicommutation, $\mathrm{End}_K A \otimes \mathbb{Q}_\ell$ is the commutant of $\rho_\ell(\mathbb{Q}_\ell[\pi])$ in $\mathrm{End}\, V_\ell(A)$, which implies (ii) of 1.1 – cf.2.4 above.

But 4.2 is much more precise: It says that, for almost all ℓ, $\rho_\ell(\mathbb{Z}_\ell[\pi])$ is exactly the commutant of $\mathrm{End}_K(A)$ in $\mathrm{End}_{\mathbb{Z}_\ell}(T_\ell(A))$!

<u>Proof of 4.2</u>: All we have to show is the last-mentioned equality of $\rho_\ell(\mathbb{Z}_\ell[\pi])$ and $\mathrm{End}_K(A)^\circ$, for almost all ℓ. We proceed by a reduction very much reminiscent of 2.4.

<u>(4.3)</u> *It suffices to show that, for almost all prime numbers* ℓ, *if* W *is a* π*-invariant subspace of the* $\mathbb{F}_\ell$*-vector space* $A[\ell](\bar{K})$, *then there is* $u \in \mathrm{End}_K A$ *such that* $W = A[\ell](\bar{K}) \cap \ker(u)$.

In fact, assuming the condition of 4.3, one immediately gets the semi-simplicity of the π-action on the $\mathbb{F}_\ell$- vector space $A[\ell](\bar{K})$. So, the algebra F_ℓ generated by the elements of

π in $\mathrm{End}_{\mathbb{F}_\ell}(A[\ell](\overline{K}))$ is a semi-simple $\mathbb{F}_\ell$-algebra. Thus, letting

$$E_\ell = \mathrm{End}_K A \otimes_{\mathbb{Z}} \mathbb{Z}/\ell\mathbb{Z} \subset \mathrm{End}_{\mathbb{F}_\ell}(A[\ell](\overline{K})) \quad ,$$

and denoting commutants by $\circ$, the theorem of bicommutation tells us that $F_\ell = E_\ell{}^\circ$ if and only if $F_\ell{}^\circ = E_\ell$. But the condition of 4.3 for $A \times A$ implies $F_\ell{}^\circ = E_\ell$, by exactly the same argument as in 2.4. So, we have $F_\ell = E_\ell{}^\circ$, for almost all primes ℓ. Finally, calling $\mathrm{End}_K A^\circ$ the commutant of $\mathrm{End}_K A$ in $\mathrm{End}_{\mathbb{Z}_\ell}(T_\ell(A))$, we have mappings

$$F_\ell \xrightarrow{\rho_\ell \otimes \mathbb{Z}/\ell\mathbb{Z}} \mathrm{End}_K A^\circ/\ell.\mathrm{End}_K A^\circ \hookrightarrow E_\ell{}^\circ \quad .$$

So, by Nakayama's lemma, $F_\ell = E_\ell{}^\circ$ implies $\rho_\ell(\mathbb{Z}_\ell[\pi]) = \mathrm{End}_K A^\circ$. This proves 4.3.

In order to prove 4.2, we have to use a result which will only be established in the following article:

<u>4.4 Theorem (see [Wüst], 3.5).</u> *For* A *with semi-stable reduction over* K, *there is a finite set of primes* T *such that, for any isogeny* $A \to B$ *over* K *of degree prime to all* $\ell' \in T$, *one has*

$$h(A) = h(B) \quad .$$

Like in 2.2, 2.3, we have to prove 4.2 only for semi-stable A. Suppose then that the condition of 4.3 fails to be true. Then there is an infinite set M of prime numbers such that for all $\ell \in M$ there is a π-invariant subspace $W_\ell \subset A[\ell](\overline{K})$ which does not come from an endomorphism u as required in

4.3. Then 4.4 and 2.8 imply that there is an infinite subset $M_\circ \subset M$ such that for all $\ell, \ell' \subset M_\circ$, $A/W_\ell \cong A/W_{\ell'}$. Taking $\ell \neq \ell'$ in $M_\circ$, call f the composite map

$$A \longrightarrow A/W_\ell \xrightarrow{\cong} A/W_{\ell'} \longrightarrow A \quad .$$

Since the degree of the last map is a power of ℓ' , the endomorphism $f \in \mathrm{End}_K A$ satisfies indeed

$$W_\ell = A[\ell](\bar{K}) \cap \ker (f),$$

contradicting our initial assumption on M . This proves 4.2.

(4.5) To conclude, let us recall (cf. [T1] and [F1]) that we could have used the weaker diophantine result on *principally polarized* abelian varieties,[F2],II 4.3, instead of 2.8, in the proof of Theorem 1.1, at the expense of working a little harder on the reduction steps of § 2. Refining 2.4, we would have had to reduce to showing that any *maximal isotropic* subspace $W \subset V_\ell(A)$ - with respect to the ℓ-adic Riemann form of some fixed principal polarization on A - is the image of some global endomorphism. This is done by an argument quite similar to the one we had to use here in the proof of 3.8 in order to get 2.8. See [Z4], 2.6, for this reduction. Incidentally, in this approach, it is legal to assume A principally polarized because, over a field extension (see 2.2) A is isogenous to some principally polarized abelian variety B ; and 1.1 is invariant under isogeny, thanks to 2.1, because isogenous varieties have isomorphic π-representations V_ℓ .

References

[Bou] N. Bourbaki, Algèbre, chap. 8; Paris 1958.

[BoL] N. Bourbaki, Groupes et algèbres de Lie, chap. 1 and chap. 2 et 3 ; Paris 1971/72.

[Deu] M. Deuring, Die Typen der Multiplikatorenringe elliptischer Funktionenkörper; Abh. Math. Sem. Hansische Univ. 14 (1941), 197-272.

[F1] G. Faltings, Endlichkeitssätze für abelsche Varietäten über Zahlkörpern; Inventiones Math. 73 (1983), 349-366.

[F2] G. Faltings, contribution to this volume (chap. I,II,VI).

[De] P. Deligne, Preuves des conjectures de Tate et de Shafarevitch; Sém. Bourbaki n°616 (1983/84).

[EGA II] A. Grothendieck, Éléments de Géometrie Algébrique, II; Publ. Math. I.H.E.S. 8 (1961).

[EGA III] A. Grothendieck, Éléments de Géometrie Algébrique, III; Publ. Math. I.H.E.S. 11 (1961).

[Gro] A. Grothendieck, Groupes de type multiplicatif: Homomorphismes dans un schéma en groupes; in: SGA 3/Schémas en groupes II, Springer Lect. Notes Math. 152 (1970).

[Groth] A. Grothendieck, Modèles de Néron et Monodromie; exp. IX in: SGA 7 I, Springer Lect. Notes Math. 288 (1972).

[Grun] F. Grunewald, contribution to this volume (chap. III).

[Hum] J.E. Humphreys, Linear algebraic groups; Springer GTM 21, 1975.

[Mar] J. Martinet, Character Theory and Artin L-functions; in: Algebraic Number fields (A. Fröhlich, ed.), Proc. LMS Symp. Durham; Acad. Press 1977.

[Mil] J.S. Milne, Etale Cohomology; Princeton U Press, 1980.

[Mu1] D. Mumford, Abelian Varieties; Oxford U Press, 1974.

[Mu2] D. Mumford and J. Fogarty, Geometric Invariant Theory (2nd enlarged edition); Springer Ergebnisse 34 (1982).

[Ri] K.A. Ribet, Twists of Modular Forms and Endomorphisms of Abelian Varieties; Math. Ann. 253 (1980), 43-62.

[Se] J.P. Serre, Abelian ℓ-adic representations and elliptic curves ; Benjamin 1968.

[Shim] G. Shimura, On the zeta-function of an abelian variety with complex multiplication; Ann. Math. 94 (1971), 504-533.

[ST] J.P. Serre and J. Tate, Good reduction of abelian varieties; Ann. Math. 88 (1968), 492-517.

[T1] J. Tate, Endomorphisms of abelian varieties over finite fields; Inventiones Math. 2 (1966), 134-144.

[T2] J. Tate, p-divisible groups; in : Proc. of a conference on *Local Fields* (Driebergen), Springer 1967.

[T3] J. Tate, Algebraic cycles and poles of zeta functions; in: Arithmetical algebraic geometry, New York (Harper & Row) 1966.

[Wüst] G. Wüstholz, contribution to this volume (chap. V).

[Z1] Ju.G. Zarhin, Isogenies of abelian varieties over fields of finite characteristic, Mat. Sb. 95(137) (1974), 461-470 = Math. USSR Sb. 24 (1974), 451-461.

[Z2] Ju.G. Zarhin, A finiteness theorem for isogenies of abelian varieties over function fields of finite characteristic; Funct. Anal. i ego Prilozh. 8 (1974), 31-34.

[Z3] Ju.G. Zarhin, A remark on endomorphisms of abelian varieties over function fields of finite characteristic; Izv. Akad. Nauk SSR, Ser. Mat. 38 (1974) = Math. USSR Izvest. 8 (1974), n°3, 477-480.

[Z4] Ju.G. Zarhin, Endomorphisms of abelian varieties over fields of finite characteristic; Izv. Akad. Nauk SSR, Ser. Mat. 39 (1975) = Math. USSR Izvest. 9 (1975), n°2, 255-260.

[Z5] Ju.G. Zarhin, Abelian varieties in characteristic p ; Mat. Zametki 19, 3 (1976), 393-400 = Math. Notes 19 (1976), 240-244.

[ZZ] H. Pohlmann, Algebraic cycles on abelian varieties of complex multiplication type; Annals of Math. 88 (1968), 161-180.

V

THE FINITENESS THEOREMS OF FALTINGS

G. Wüstholz

Contents:

§1 Introduction

In this chapter we shall state the finiteness theorems of Faltings and give very detailed proofs of these results. In the second section we shall beginn with the finiteness theorem for isogeny classes of abelian varieties with good reduction outside a given set of primes. Here we use in an essential way the Tate conjecture which is proved in much detail in [Sch].

In the third section we give then the proof of the finiteness theorem for isomorphism classes of abelian varieties with prescribed good reduction. Here we use deeply the results of Raynaud on finite group schemes of typ $(p,\dots,p)$. Again very detailed proofs of the results which are used are given in [Gru] .

In section 4 we shall use the results of the preceeding two sections in order to give Faltings' proof of the Mordell conjecture. Here we use the construction of Parshin [Pa] which associates to a rational point a certain curve with good reduction outside of a finite set of primes which does not depend on the point. This makes it possible to apply the finiteness theorem on isomorphism classes.

In the last section we give then a proof of Siegel's theorem on the finiteness of integer points on curves. This proof does not use diophantine approximations.

This paper profited very much from the Exposés given by Deligne [De] and Szpriro [Sz] in the Séminaire Bourbaki.

§2 The finiteness theorem for isogeny classes

2.1 *Eigenvalues of the Frobenius automorphism* .

Let K be an algebraic number field, S a finite set of finite places of K and ℓ a prime number. Suppose that A is an abelian variety defined over K with good reduction outside of S . Let further v be a finite place of K not in S and not deviding ℓ , F_v the Frobenius automorphism at the place v acting on the Tate module $T_\ell(A)$ of A. Then we can define the characteristic polynomial $P_h(T)$ for $0 \leq h \leq 2g$ $(g = \dim A)$ by

$$P_h(T) = \det(T\cdot id - F_v \mid \overset{h}{\wedge} T_\ell(A)) .$$

If we denote by N_v the number of elements of the residue field K_v of v , then the following theorem is a consequence of a result of Weil.

Theorem 2.1. *For* $0 \leq h \leq 2g$ *the polynomials* $P_h(T)$ *have integer coefficients and do not depend on* ℓ . *Furthermore their complex zeroes have absolute values equal to* $N_v^{h/2}$.

We shall use this result later on in a modified form. For this let p be a prime number (replacing v) not deviding ℓ and

$$\pi = \mathrm{Gal}\ (\overline{K}/K) ,$$
$$\tilde{\pi} = \mathrm{Gal}\ (\overline{\mathbb{Q}}/\mathbb{Q}) .$$

Then denoting by $\mathrm{Res}_{K/\mathbb{Q}} A$ the Weil restriction of A (see [We]) we obtain an abelian variety defined over $\mathbb{Q}$ that has good reduction outside the set of primes which are divisible by the places contained in S or ramify in K. Furthermore we have

$$T_\ell(\mathrm{Res}_{K/\mathbb{Q}} A) = \mathrm{Ind}_{\tilde{\pi}}^{\pi} (T_\ell(A)).$$

Now we can apply Theorem 2.1 to $\mathrm{Res}_{K/\mathbb{Q}} A$ and F_p. Here we have

$$\dim \ \mathrm{Res}_{K/\mathbb{Q}} A = [K:\mathbb{Q}] \cdot \dim A$$

and therefore we obtain the following corollary.

Corollary 2.2. *For* $0 \leq h \leq 2[K:\mathbb{Q}] \cdot g$ *the polynomials*

$$P_h(T) = \det (T\cdot \mathrm{id} - F_p \mid \overset{h}{\wedge} \mathrm{Ind}_{\tilde{\pi}}^{\pi} (T_\ell(A)))$$

have integer coefficients and do not depend on ℓ. *The absolute values of their complex zeroes are equal to* $p^{h/2}$.

2.2 The density Theorem of Čebotarev.

Let K, S, ℓ be as before, Σ_K the set of all finite places of K and S_ℓ the set of finite places of K consisting of S and those places which divide ℓ. Now let P be a subset of Σ_K and for each integer n let $a_n(P)$ be the number of

$v \in P$ with $N_v \leqq n$, where N_v is equal to the number of elements of the residue field k_v of v.

Then one says that P has density $\alpha(P)$ if

$$\alpha(P) = \lim_{n\to\infty} a_n(P)/a_n(\Sigma_K)$$

exists. Since by the prime number theorem

$$a_n(\Sigma_K) \sim n/\log n$$

one gets

$$a_n(P) = \alpha(P)\cdot n/\log n + o(n/\log n).$$

Now we can state the density theorem of Čebotarev (see [Se 1]).

Theorem 2.3. *Let* L *be a finite Galois extension of the number field* K *with Galois group* G *. Let* X *be a subset of* G *that is stable under conjugation. Denote by* P_X *the set of places* $v \in \Sigma_K$ *unramified in* L *such that the conjugacy class of the Frobenius automorphism* F_v *is contained in* X *. Then*

$$\alpha(P_X) = |X|/|G|.$$

We shall use later on the following version of the density theorem.

Corollary 2.4. *Let* $K' \supseteq K$ *be a finite Galois extension of* K *unramified outside of* S_ℓ *. Then there exists a finite set of places* T *of* K *such that* $T \cap S_\ell = \emptyset$ *and the conjugacy classes of the Frobenius automorphisms* $F_v (v \in T)$ *cover all of* $Gal(K'/K)$.

Remark. For effective versions of the density theorem of Čebotarev see [Se 2] .

2.3 The Theorem of Hermite-Minkowski

Let K be as before an algebraic number field and $S \subset \sum_K$ a finite set of finite places of K. Then we have the following well-known result of Hermite-Minkowski.

Theorem 2.6. *There exist only finitely many Galois extensions* $L \supseteq K$ *unramified outside of* S *and of degree at most equal to a given number* d.

Sketch of the proof. The set S of finite places determines the prime factors of the discriminants of the extensions which are unramified outside of S . It remains to bound the exponents of these prime factors.

This is done by well-known estimates of the exponent in terms of the ramification indices at the places in question. These can be bounded in terms of the degree and consequently by d. Therefore there are only finitely many possibilities for the discriminant.

Remark. The number of such extensions $L \supseteq K$ which are unramified outside of S and of degree at most equal to d can be effectively determined (see [Se 2]).

2.4 The finiteness theorem for isogeny classes.

Let K be an algebraic number field, S a finite set of finite places and ℓ a prime number.

Lemma 2.6. *Let A/K be an abelian variety defined over K with good reduction outside of S. Then for any fixed place $v \notin S$ prime to ℓ there are only finitely many possibilities for the local L-factor*

$$L_v(A,s) = \det\,(1 - N_v^{-s}F_v \mid T_\ell(A))^{-1} .$$

Proof. Consider the polynomial

$$P(T) = \det\,(T\cdot \mathrm{id} - F_v \mid T_\ell(A)).$$

By Theorem 2.1 this polynomial has integer coefficients and its zeroes have absolute values equal to $N_v^{1/2}$. Hence there are only finitely many possibilities for the coefficients and the number of the polynomials $P(T)$ is bounded. Now the statement follows directly.

Remark. This number of possibilities for the local L-factor can be effectively determined in terms of N_v and $g = \dim A$.

We denote by $\tilde{K}$ a finite Galois extension of K which contains all Galois extensions $K' \supseteq K$ with

$$[K':K] < \ell^{2d^2}$$

and which are unramified outside of S. Let T be the set of finite places constructed in Corollary 2.4 for S and $L = \tilde{K}$, d a positive integer.

Proposition 2.7. *Let* $\rho_1, \rho_2 : \mathrm{Gal}(\overline{K}/K) \longrightarrow GL(d, \mathbb{Q}_\ell)$ *be two semi-simple representations with*

$$\mathrm{Trace}\ \rho_1(F_v) = \mathrm{Trace}\ \rho_2(F_v) \quad (v \in T)$$

which are unramified outside of S*. Then* ρ_1 *and* ρ_2 *are isomorphic.*

Proof. It is a well-known fact in representation theory that semi-simple representations are isomorphic if their traces are equal. In order to prove that ρ_1 and ρ_2 are isomorphic it suffices therefore to show that

$$\mathrm{Trace}\ \rho_1(\sigma) = \mathrm{Trace}\ \rho_2(\sigma)$$

for all $\sigma \in \mathrm{Gal}(\overline{K}/K)$.

Consider the image M of the algebra $\mathbb{Z}_\ell[\mathrm{Gal}(\overline{K}/K)]$ under the homomorphism

$$\rho_1 \times \rho_2 : \mathbb{Z}_\ell[\mathrm{Gal}(\bar{K}/K)] \longrightarrow M_d(\mathbb{Q}_\ell) \times M_d(\mathbb{Q}_\ell).$$

and define the function $f : M \longrightarrow \mathbb{Q}_\ell$ by

$$f(m,m') = \mathrm{Trace}\ m - \mathrm{Trace}\ m'$$

for $(m,m') \in M$. We have to show that f is identically zero. For this it suffices to show that f vanishes on a set of generators. We shall show that M is generated over $\mathbb{Z}_\ell$ by the images of the conjugacy classes of the Frobenius automorphisms F_v for $v \in T$. Since

$$f(\rho_1(F_v), \rho_2(F_v)) = 0 \qquad (v \in T)$$

by hypothesis the function will then be identically zero. In order to show that the images of the conjugacy classes of the $F_\nu (\nu \in T)$ generate the module M over $\mathbb{Z}_\ell$ it suffices to show that they generate $M/\ell M$. This follows from the Lemma of Nakayama since $\mathbb{Z}_\ell$ is local and M finitely generated.

The representation $\rho = \rho_1 \times \rho_2$ induces a representation

$$\bar{\rho}: \mathrm{Gal}(\bar{K}/K) \longrightarrow (M/\ell M)^*$$

of the Galois group $\mathrm{Gal}\ (\bar{K}/K)$ into the group of units in $M/\ell M$ and the image of $\bar{\rho}$ generates $M/\ell M$. Since

$$\# \ (M/\ell M)^{*} < \ell^{2d^2}$$

and since the representations are unramified outside of S the representation ρ factorizes over $Gal(\tilde{K}/K)$. The conjugacy classes of the Frobenius automorphisms F_v for $v \in T$ cover $Gal(\tilde{K}/K)$ by construction so that their images under $\bar{\rho}$ generate $M/\ell M$ over $\mathbb{Z}_\ell$. This completes the proof of the Proposition.

We are now able to prove the Main Theorem of this section. We use two more facts which are proved in [Sch] , namely

1. the action of $\pi = Gal(\bar{K}/K)$ on

$$V_\ell(A) := T_\ell(A) \otimes_{\mathbb{Z}_\ell} \mathbb{Q}_\ell$$

is semi-simple (Theorem 1.1 in [Sch]),

2. two abelian varieties A, A' both defined over K are isogenous over K if and only if the π- Modules $V_\ell(A)$ and $V_\ell(A')$ are isomorphic (Cor. 1.3 in [Sch]).

To an abelian variety A defined over K one associates an L-series in the following way. For $v \in \sum_K$ let I_v be the inertia subgroup of $\pi = Gal(\bar{K}/K)$ and $T_\ell(A)^{I_v}$ the fixed part under the action on I_v of $T_\ell(A)$. Then the action of $F_v \in \pi/I_v$ is well-defined on $T_\ell(A)^{I_v}$ and we put

$$L_v(A,s) = \frac{1}{\det(id - N_v^{-s} F_v \mid T_\ell(A)^{I_v})}$$

Note that for $v \notin S_\ell$ one has $T_\ell(A) = T_\ell(A)^{I_v}$, where S is the set of places $v \in \Sigma_K$ where A has bad reduction and S_ℓ is defined as in 2.2. Then $L(A,s)$ is defined as

$$L(A,s) = \prod_{v \in \Sigma_K} L_v(A,s) \quad .$$

This function is defined for $\mathrm{Re}\, s > 3/2$.

Theorem 2.8. *Let* S *be a finite set of finite places of* K , $g \geqq 1$ *an integer. Then there exist only finitely many isogeny classes of abelian varieties defined over* K *of dimension* g *and with good reduction outside of* S .

Proof. We call two such abelian varieties A, A' equivalent if for all $v \in T$

$$L_v(A,s) = L_v(A',s) \quad .$$

Here T is defined as in Proposition 2.7. (It would indeed suffice to call A, A' equivalent if the traces of the Frobenius are equal). Then we deduce from Lemma 2.6 that there are only finitely many equivalence classes. We proceed to show that equivalent abelian varieties are isogeneous. Since two abelian varieties A and A' as above are isogeneous if and only if the π-modules $V_\ell(A)$ and $V_\ell(A')$ are isomorphic (this is 2. above) we need only to show that A and A' are equivalent if and only if $V_\ell(A)$ and $V_\ell(A')$ are isomorphic as π-modules.

Suppose first that A and A' are equivalent. Then the π-modules $V_\ell(A)$ and $V_\ell(A')$ correspond to representations

$$\rho,\rho' : \mathrm{Gal}(\overline{K}/K) \longrightarrow GL(2g,\mathbb{Q}_\ell) \quad .$$

These representations are semi-simple and unramified outside of S and satisfy

$$L_v(A,s) = L_v(A',s) \qquad (v \in T) \ .$$

From this it follows that

$$\mathrm{Trace}\ \rho(F_v) = \mathrm{Trace}\ \rho'(F_v)$$

for all $v \in T$. By Proposition 2.7 the representations ρ and ρ' are isomorphic and therefore $V_\ell(A)$ and $V_\ell(A')$ are isomorphic as π-modules.

Next suppose that $V_\ell(A)$ and $V_\ell(A')$ are isomorphic π-modules. Then the corresponding representations ρ and ρ' are isomorphic and therefore

$$L_v(A,s) = L_v(A',s)$$

for all v, i.e. A and A' are equivalent. It follows that the number of isogeny classes is equal to the number of equivalence classes. Since this number is finite the theorem follows.

Remark. Theorem 2.8 is completely effective: it is possible to establish an upper bound for the number of isogeny classes effectively in terms of S, g, ℓ and K .

§3 The finiteness theorem for isomorphism classes

3.1 *Statement of the theorem and first reductions*

Let K,S,g be as before and $d > 0$ an integer, R the ring of integers of K. The following theorem was conjectured by Shafarevich.

Theorem 3.1. *There are only finitely many isomorphism classes of* d*-fold polarized abelian varieties defined over* K *of dimension* g *with good reduction outside of* S.

Remark. It can be shown that this remains true even without polarisation. We shall establish in this section a Theorem (Theorem 3.5) and show how Theorem 3.1 will follow from this result. It will be proved then in the next section. But first we shall make some simple reductions.

Reductions: 1. Without loss of generality we can assume that $d = 1$, i.e the abelian varieties are principally polarized. This is obtained with Zarhin's trick (see Lemma 3.8 of [Sch]).

2. Because of Theorem 2.8 it suffices to prove that there are only by finitely many isomorphism classes within a given isogeny class. Let A be a prinicpally polarized abelian variety defined over K, of dimension g and with good reduction outside of S. Then we denote by $cl(A)$ the isogeny class of A.

3. We may assume without loss of generality that all B in $cl(A)$ can be extended to semi-abelian varieties over spec R. This can be obtained by a finite galois extension $K' \supseteq K$ for which the torsion points of order 4 and 3 of A become K'-rational. We replace then K by K', R by R', the integral closure of R in K', and A by $A \times_K K'$. This is the theorem on semi-stable reduction (see [SGA 7.I], Exposé IX). Then if $B \in cl(A)$ and A is semi-stable the abelian variety B is also semi-stable.

3.2 *Some auxiliary results*

Let N be a finite set of prime numbers, A a principally polarized abelian variety defined over K, ℓ as usual. Then we denote by $A[\ell^n]$ for integers $n \geqq 0$ the set of torsion points of A with order dividing ℓ^n.

Lemma 3.2. *Suppose that* A' *is isogeneous to* A *and*

$$T_\ell(A') \cong T_\ell(A)$$

for all $\ell \in N$ *as* π*-modules. Then there exists an isogeny*

$$\varphi : A' \longrightarrow A$$

of degree prime to all ℓ *in* N.

Remark. If the degree $\deg\varphi$ of φ is prime to each ℓ in N we shall denote this by $(\deg \varphi, N) = 1$.

Proof. Since $\mathrm{Hom}(A',A)$ is dense in

$$\prod_{\ell \in N} \mathrm{Hom}_{\pi} (T_{\ell}(A'), T_{\ell}(A))$$

it follows (see Theorem 4.1 of [Sch]) that

$$\mathrm{Hom}\,(A,A') \otimes \prod_{\ell \in N} \mathbb{Z}_{\ell} \cong \prod_{\ell \in N} \mathrm{Hom}_{\pi} (T_{\ell}(A'), T_{\ell}(A)).$$

Let now φ_{ℓ} for $\ell \in N$ be the given isomorphisms

$$\varphi_{\ell} : T_{\ell}(A') \longrightarrow T_{\ell}(A) \quad .$$

Then there exist $\Psi_1 \in \mathrm{Hom}\,(A',A)$ and $\Psi_2 \in \mathrm{Hom}\,(A',A) \otimes \mathbb{Z}_{\ell}$ such that

$$\Psi = \Psi_1 + \ell \Psi_2$$

satisfies

$$T_{\ell}(\Psi) = \varphi_{\ell}$$

for $\ell \in N$. From this we deduce that the kernel of Ψ and hence that of Ψ_1 is finite. It follows that Ψ_1 is an isogeny and it remains to show that

$$(\deg \Psi_1, N) = 1 .$$

Suppose that the prime number ℓ divides

$$(\deg \Psi_1, N) .$$

Then there exists an element x in A' of order ℓ such that $\Psi_1(x) = 0$. Hence

$$\Psi(x) = \Psi_1(x) + \ell\Psi_2(x) = 0$$

and therefore

$$\varphi_\ell(x) = T_\ell(\Psi)(x) = 0 .$$

But φ_ℓ is an isomorphism and we may conclude that $x = 0$. Since we have assumed that the order of x is equal to ℓ we have obtained a contradiction. It follows that

$$(\deg \Psi_1, N) = 1$$

as claimed. This concludes the proof of Lemma 3.2.

The next Lemma is an important step towards the proof of Theorem 3.1.

<u>Lemma 3.3.</u> *There are only finitely many isomorphism classes of* $\mathbb{Z}_\ell[\pi]$-*invariant lattices in* $T_\ell(A)\otimes_{\mathbb{Z}_\ell}\mathbb{Q}_\ell$.

Proof. This follows directly from the Jordan-Zassenhaus Theorem (for a proof see [Rei]). For if M_ℓ denotes the $\mathbb{Z}_\ell$-subalgebra generated in $\mathrm{End}_{\mathbb{Z}_\ell}(T_\ell(A))$ by π then the algebra $M_\ell \otimes_{\mathbb{Z}_\ell} \mathbb{Q}_\ell$ is semi-simple by Theorem 1.1 in [Sch] .

3.3 Heights on isogeny classes

Let N denote again a non-empty set of prime numbers, A a principally polarized abelian variety defined over K and $cl(A)$ the isogeny class of A .

Proposition 3.4. *There exist an integer* $n \geq 1$ *depending only on* N *and* $A_1, \ldots, A_n \in cl(A)$ *with the following property. Let* $B \in cl(A)$ *be any abelian variety isogeneous to* A *. Then there exists an integer* $i = i(B)$ *with* $1 \leq i \leq n$ *and an isogeny*

$$\varphi : B \longrightarrow A_i$$

with $(\deg \varphi, N) = 1$.

Proof. According to Lemma 3.3 there exist an integer n depending only on N and $A_1, \ldots, A_n \in cl(A)$ with the following property.

Let $B \in cl(A)$ be any abelian variety isogeneous to A . Then there exists an integer $i = i(B)$ with $1 \leq i \leq n$ such that

$$T_\ell(B) \cong T_\ell(A_i)$$

for all $\ell \in N$ as π-modules. By Lemma 3.2 it follows that there exists an isogeny

$$\varphi: B \longrightarrow A_i$$

with $(\deg \varphi, N) = 1$. This proves the Proposition.

Now we come to the main step in the proof of Theorem 3.1. This is the following theorem.

<u>Theorem 3.5</u> *Let* A *be a principally polarized abelian variety defined over* K *with semi-stable reduction. Then there exists a finite set* N *of primes with the following property. Let* $\varphi: A' \longrightarrow A$ *be an isogeny with* $(\deg \varphi, N) = 1$ *. Then one has*

$$h(A') = h(A) \quad .$$

<u>Corollary 3.6.</u> *One has*

$$h(cl(A)) = \{h(A_1), \ldots, h(A_n)\}$$

if $A_1, \ldots, A_n$ *are the abelian varieties constructed in Proposition* 3.4 *for the* N *given by Theorem* 3. 5.

Proof. Obvious.

Corollary 3.7. *There exists a constant* $c > 0$ *depending on* $A_1, \ldots, A_n$ *(as in Corollary* 3.6*) such that for* $B \in cl(A)$ *one has*

$$h(B) \leq c .$$

Proof. Obvious.

Remark. The constant c can be effectively determined in terms of $K, N, h(A_1), \ldots, h(A_n)$. But it is not possible to give an effective bound for the $h(A_i)$ $(1 \leq i \leq n)$.

We shall prove Theorem 3.5 in the next section. Since Theorem 3.1 follows very easily from Theorem 3.5 we shall give the proof now. For this we need the following result which is proved in [Fa II] , Theorem 4.3.

Theorem 3.8. *Let* K *be a number field. Fix an integer* $g \geq 1$ *and a real number* $c > 0$ *. Then there are up to isomorphism only finitely many principally polarized semistable abelian varieties* A *over* K *of dimension* g *such that* $h(A) \leq c$.

Proof of Theorem 3.1. From Theorem 2.8 it follows that the number of isogeny classes is bounded. In each isogeny class the height is bounded by Corollary 3.7. By Theorem 3.8 there are only finitely many isomorphism classes of principally polarized abelian varieties of bounded height. Together with reduction 1 and reduction 3 Theorem 3.1 follows now.

3.4 *Galois representations and the Theorem of Raynaud*

The arguments in this section are very similar to those used in [Sch] , 3.6. Since we are working here mod ℓ, the ℓ-divisible groups are replaced by finite group schemes.

Let K be a number field as usual, R its ring of integers and $m = [K:\mathbb{Q}]$, $\pi = \mathrm{Gal}(\overline{K}/K)$ and $\tilde{\pi} = \mathrm{Gal}(\overline{\mathbb{Q}}/\mathbb{Q})$. We fix some prime number ℓ and denote by $\mathbb{F}_\ell$ the finite field with ℓ elements. Let V be a finite dimensional $\mathbb{F}_\ell$-module with dim V = h. Suppose that

$$\rho:\pi \longrightarrow GL(V)$$

is a Galois representation. Then the module V becomes a π-module and we can define the module

$$\tilde{V} = \mathrm{Ind}_{\pi}^{\tilde{\pi}} V \quad .$$

This is a $\tilde{\pi}$-module and to it corresponds the representation

$$\tilde{\rho}:\tilde{\pi} \longrightarrow GL(\tilde{V})$$

where

$$\tilde{\rho} = \mathrm{Ind}_{\pi}^{\tilde{\pi}} \rho \quad .$$

To each such representation we can associate the

one-dimensional determinant representation. It follows that we obtain two further representations

$$\det \rho : \pi \longrightarrow \mathbb{F}_\ell$$

and

$$\det \tilde{\rho} : \tilde{\pi} \longrightarrow \mathbb{F}_\ell \quad .$$

Both induce representations of π^{ab} and $\tilde{\pi}^{ab}$ of the abelianized Galois groups. We denote by

$$\mathrm{Ver}_{\tilde{\pi}}^{\pi} : \pi \longrightarrow \pi^{ab}$$

the canonical projection $\tilde{\pi} \longrightarrow \tilde{\pi}^{ab}$ followed by the transfer map $\tilde{\pi}^{ab} \longrightarrow \pi^{ab}$. We put

$$\tilde{\chi} := \det \tilde{\rho} : \tilde{\pi} \longrightarrow \mathbb{F}_\ell$$

and

$$\chi := (\det \rho) \circ (\mathrm{Ver}_{\tilde{\pi}}^{\pi}) : \tilde{\pi} \longrightarrow \mathbb{F}_\ell \quad .$$

χ and $\tilde{\chi}$ are characters of $\tilde{\pi}$ with values in $\mathbb{F}_\ell$. If we denote by

$$\varepsilon : \tilde{\pi} \longrightarrow \{\pm 1\}$$

the signature of the permutations induced by the elements of $\tilde{\pi}$ on $\tilde{\pi}/\pi$ then χ and $\tilde{\chi}$ are linked as follows.

Proposition 3.9. *We have*

$$\tilde{\chi} = \varepsilon^h \chi \quad .$$

Proof. See [Ma], Proposition 3.2.

We apply this in the following situation. Let A be an abelian variety as usual (principally polarized, defined over K, semistable reduction over K),

$$\varphi: A' \longrightarrow A$$

an isogeny with kernel G that is annihilated by ℓ and $\deg \varphi = \ell^h$. Then G is a quasi-finite and flat group scheme over R. If v is a place of K with $v|\ell$ and if A has good reduction at v then

$$G_v = G \otimes R_v$$

is a finite and flat group scheme over R_v. Associated to A' and G are several modules over $\mathbb{F}_\ell$, namely

$$V_\ell = T_\ell(A')/\ell T_\ell(A')$$

and

$$W_\ell = G(\bar{K}) \subseteq V_\ell .$$

Then both V_ℓ and W_ℓ are π-modules and we can apply the foregoing to $V = W_\ell$. For this we put

$$\tilde{V}_\ell = \mathrm{Ind}_{\pi}^{\tilde{\pi}} V_\ell$$

and

$$\tilde{W} = \mathrm{Ind}_{\pi}^{\tilde{\pi}} W_\ell$$

and these are both $\tilde{\pi}$-modules. We have then the representation $\rho : \pi \longrightarrow GL(W_\ell)$ and the induced representation $\tilde{\rho} = \mathrm{Ind}_{\pi}^{\tilde{\pi}} \rho : \tilde{\pi} \longrightarrow GL(\tilde{W}_\ell)$. We have further associated to ρ and $\tilde{\rho}$ the characters $\chi = \det \rho \circ \mathrm{Ver}_{\tilde{\pi}}^{\pi}$ and $\tilde{\chi} = \det \tilde{\rho}$. Since $h = \dim W_\ell$ we obtain from Proposition 3.9 that

$$\tilde{\chi} = \varepsilon^h \chi .$$

Let

$$\chi_0 : \tilde{\pi} \longrightarrow \mathbf{Z}_\ell^*$$

be the cyclotomic character and by abuse of notation we denote also by χ_0 its reduction mod ℓ ,

$$\chi_0 : \tilde{\pi} \longrightarrow \mathbf{F}_\ell^* ,$$

i.e. the compositum with the canonical projection $\mathbb{Z}_\ell \longrightarrow \mathbb{F}_\ell$.

Lemma 3.10. *The character* χ *is a power of* χ_0.

Proof. Since A' has semi-stable reduction outside of $v|\ell$ it follows that ρ operates unipotently ([SGA 7]) on I_w, $w \nmid \ell$. Hence χ is unramified outside of ℓ . But then it follows that χ is a power of the character χ_0 (using the theorem of Kronecker-Weber and the fact that $\mathbb{Q}$ does not possess any unramified extensions). This proves the Lemma.

We are going now to compute the exact power of χ_0 . This is done using a result of M. Raynaud on finite group schemes.

Let as before G be the kernel of an isogeny $\phi : A' \longrightarrow A$ annihilated by ℓ and assume that for each place v dividing ℓ

(i) A' has good reduction at v,

(ii) K is unramified at v.

Let $\Omega^1_{G/R}$ be the module of differentials of G , $s : R \longrightarrow A$ the zero section and the non-negative integer d defined by

$$\ell^d = \# s^*(\Omega^1_{G/R}) \quad .$$

Then d satisfies (see [Gru],Proposition 2.7)

$$0 \leq d \leq m \cdot g \qquad\qquad (m = [K:\mathbb{Q}], g = \dim A').$$

We have then the following theorem.

Theorem 3.11. *One has*

$$\tilde{\chi} = \varepsilon^h \chi_0^d \quad .$$

Proof. See [Gru], Theorem 4.6.

Remark. The condition (ii) on the place v implies that the ramification index e_v is equal to 1 and this implies that $e_v \leqq \ell - 1$ for every v dividing ℓ. The condition (i) implies that G_v is a finite and flat group scheme over R_v.

3.5. *Proof of Theorem* 3.5

Let A', K be the same as in the preceeding section, p and ℓ two different prime number such that K is unramified at p and ℓ, and for each place v of K such that $v|p\cdot\ell$ the abelian variety A' has good reduction at v. Let F_p be the Frobenius automorphism at the prime p and for each integer h with $1 \leqq h \leqq 2mg$ define the polynomials $P_h(T)$ as

$$P_h(T) = \det(T\cdot id - F_p \mid \overset{h}{\wedge} \operatorname{Ind}_{\pi}^{\tilde{\pi}} T_\ell(A')) .$$

Then define the finite set N of primes as follows:
A prime number p' is in N if and only if one of the following conditions is satisfied:

(i) $p' = p$, $p' = 2$,

(ii) K is ramified at p' ,

(iii) for some place v of K with $v|p'$ the abelian variety has bad reduction at v ,

(iv) for $0 \leq j \leq gm$, $0 \leq h \leq 2gm$ such that $j \neq h/2$ the prime p' divides one of the numbers $P_h(\pm p^j)$ (this number is non-zero by Theorem 2.1).

This is a finite set of prime number which can be determined effectively. We sall show now that the set N has the desired property. Note that the set N does not depend on ℓ (Theorem 2.1). Therefore we may choose ℓ such that $\ell \notin N$.
Now let

$$\phi : A' \longrightarrow A$$

be an isogeny such that $(\deg \phi, N) = 1$. We may assume without loss of generality that $\deg \phi$ is a power of a prime number ℓ not contained in N . Furthermore we may assume that ℓ annihilates the kernel of ϕ. All this can be achieved by simple reductions. Let the notations be as in section 3.4.

Since $\chi_0(F_p) = p$ we obtain from Theorem 3.11

$$\tilde{\chi}(F_p) \equiv \varepsilon^h(F_p)\chi_0^d(F_p) \equiv \pm p^d$$

modulo ℓ. It follows that

$$\pm p^d$$

is a zero of the congruence

$$P_{mh}(T) \equiv 0 \mod \ell \quad .$$

Thus

$$\ell \mid P_{mh}(\pm p^d)$$

and because of the definition of N

$$d = \frac{mh}{2} \quad .$$

Now apply the height formula for isogenies ([Sch],Theorem 3.1) and obtain

$$\begin{aligned} h(A) - h(A') &= \frac{1}{2} \log (\deg\phi) - \frac{1}{[K:\mathbb{Q}]} \log (\# s^* \Omega^1_{G/R}) \\ &= \frac{h}{2} \log \ell - \frac{d}{m} \log \ell \\ &= 0 \quad . \end{aligned}$$

This proves the Theorem.

§4 Proof of Mordell's conjecture

4.1 *The theorem of Torelli.*

Let again K be a number field, S a finite set of finite places of K . Then we have the following Lemma.

Lemma 4.1. *There exists a finite extension* $K' \supseteq K$ *, such that for any abelian variety* A *over* K *of dimension* g *, with good reduction outside of* S *the abelian variety*

$$A \otimes_K K'$$

is semistable and has all its 12*-division points* K'*-rational.*

Proof. See [Fa], II, Lemma 4.2.

The general reference for the following is [Mu], chapter V,VI,VII. Let B be a noetherian scheme and $g \geqq 2$ be an integer. By a curve X of genus g over B we understand a morphism $p : X \longrightarrow B$ which is smooth, proper and whose geometric fibres are irreducible curves of genus g. Let now B any given noetherian scheme. Then we denote by $M_g(B)$ the set of curves X of genus g over B modulo isomorphisms.

By an abelian scheme A over B of dimension g we understand a group scheme $p : A \longrightarrow B$ for which p is smooth, proper and has geometrically connected fibres. For integers

$n,d \geqq 1$ denote by $A_{g,d,n}(B)$ the set of triples consisting of

(i) an abelian scheme A over B of dimension g,

(ii) a polarization of A of degree d^2,

(iii) a level n structure of A over B

up to isomorphism. If $n = d = 1$ we simply write

$$A_g(B) = A_{g,1,1}(B) \quad .$$

Then M_g and $A_{g,d,n}$ are functors which associate to a noetherian scheme a set. There exists a functor

$$j : M_g \longrightarrow A_{g,1,1}$$

which associates to a curve X over B of $M_g(B)$ its Jacobian $J(X/B)$ (see [Mu], VII 4).

From now on we let B be spec K for the number field K at the beginning and $M_g(\text{spec } K)_S$ the subset of $M_g(\text{spec } K)$ consisting of the curves X over K with good reduction outside of S . In the same way we define $A_g(\text{spec } K)_S$ as the subset of $A_g(\text{spec } K)$ consisting of the abelian varieties A over K with good reduction outside of S . Then it can be shown that the restriction $j(\text{spec } K)_S$ of $j(\text{spec } K)$ to $M_g(\text{spec } K)_S$ maps into $A_g(\text{spec } K)_S$ (use [Mu], Prop. 6.9).

Theorem 4.2. *The map*

$$j(\mathrm{spec}\ K)_S : M_g(\mathrm{spec}\ K)_S \longrightarrow A_g(\mathrm{spec}\ K)_S$$

has finite fibres and the number of elements in each fibre is uniformly bounded.

Proof. Suppose that X and Y are in $M_g(\mathrm{spec}\ K)_S$ such that

$$j(\mathrm{spec}\ K)_S(X) = j(\mathrm{spec}\ K)_S(Y) \quad .$$

Then

$$j(\mathrm{spec}\ \bar{K})_S(X\otimes\bar{K}) = j(\mathrm{spec}\ \bar{K})_S(Y\otimes\bar{K}).$$

It follows from Torelli's Theorem ([Mu],VII.4) that

$$X\otimes_K\bar{K} \cong Y\otimes_K\bar{K} \quad .$$

Let $K' \supseteq K$ be the field constructed in Lemma 4.1 and $\pi = \mathrm{Gal}(\bar{K}'/K')$. If φ denotes the above isomorphism then for $\sigma \in \pi$ one gets isomorphisms

$$\varphi^{\sigma} : X \otimes_K\bar{K}' \longrightarrow Y\otimes_K\bar{K}' \;.$$

Consider

$$\Phi(\sigma) = (\varphi^{\sigma})^{-1} \circ \varphi.$$

Then $\Phi(\sigma)$ is an automorphism of $X \otimes_K \overline{K'}$, hence of finite order (since for curves X of genus $g \geq 2$ the group $\mathrm{Aut}(X)$ is finite).

The automorphism $\Phi(\sigma)$ induces the identity on the 12-division points on the Jacobian of $X \otimes_K K'$ and therefore by the subsequent Lemma the identity on the Jacobian of $X \otimes_K K'$. It follows that $\varphi = \varphi^{\sigma}$ for $\sigma \in \pi$.

This implies that

$$X \otimes_K K' \cong Y \otimes_K K'$$

over K' . Finally the set of curves X over K which become isomorphic over K' is parametrized by a subset of the finite set

$$H^1(\mathrm{Gal}(K'/K)\ ,\ \mathrm{Aut}(X \otimes_K K'))$$

by Galois cohomology. This proves the Theorem.

In the proof of Theorem 4.2 we have used the following Lemma of Serre.

Lemma 4.3. *Let* A/K *be an abelian variety over* K . *Suppose that* $\varphi : A \longrightarrow A$ *is an endomorphism which induces the identity on the* 12-*division points of* $A(K)$. *Then* φ *is the identity on* A .

Proof. See Sém. Cartan, 1960/61, Exposé 17, ([SC]).

4.2 *The Shafarevich conjecture for curves*

Let K be a number field as in the preceeding section and S a finite set of finite places, $g \geq 2$ as integer. Then the following result was conjetured by Shafarevich.

Theorem 4.4. *There are only finitely many isomorphism classes of smooth connected curves over* K *with good reduction outside of* S .

Proof. We know by Theorem 3.1 that there are only finitely many isomorphism classes of principally polarized abelian varieties defined over K with good reduction outside of S. Now Theorem 4.4 follows from Theorem 4.2.

4.3 *Coverings*

For the proof of the Mordell conjecture we need some facts about ramified coverings of curves. In this section we give a short account of these facts.

As usual let K be an algebraic number field and R its ring of integers and S a finite set of finite places of K. We denote by U the open set

$$U = \text{spec } R - S \quad .$$

We shall also need the Hilbert class field $K' \supseteq K$ of K .

Its ring of integers is denoted by R' and S' is the set of primes of R' lying over S, $U' = \text{spec } R' - S'$. Finally we let $p : X \longrightarrow U$ be a curve (see section 4.1). The basic tool is the following construction.

Proposition 4.5. *Let* A *be a quasi-coherent sheaf of* O_X*- algebras. Then there exists an unique scheme* Y *over* U *and a morphism* $f : Y \longrightarrow X$ *over* U *such that for every open affine* $V \subseteq X$ *we have*

$$f^{-1}(V) = \text{spec } A(V) ,$$

and for every inclusion $U \hookrightarrow V$ *of open affines of* Y *the morphism*

$$f^{-1}(U) \hookrightarrow f^{-1}(V)$$

corresponds to the restriction homomorphism $A(V) \rightarrow A(U)$.

Proof. See [Ha], II, Ex. 5.17.

Remark. The scheme Y is denoted by *spec* A.

We shall also make use of the following result of Grothendieck.

Proposition 4.6. *Let* D *be an effective divisor on the generic fibre* X_K *of* X*. Then there exists an uniquely determined closed subscheme* $\tilde{D}$ *of* X *flat over* U *such that* $D = \tilde{D}_K$ *, the generic* *fibre of* $\tilde{D}$.

Proof. See Proposition 2.8.5 in [EGA IV].

Now let D be an effective divisor on X_K and $\mathcal{O}_{X_K}(-D)$ the corresponding invertible sheaf and suppose that

$$\mathcal{O}_{X_K}(-D) \cong \mathcal{L}_{X_K}^{\otimes n}$$

for some invertible sheaf $\mathcal{L}_{X_K}$ on X_K and some integer $n \geq 1$. By Proposition 4.6 the sheaves $\mathcal{O}_{X_K}(-D)$ and $\mathcal{L}_{X_K}$ extend to invertible sheaves $\mathcal{O}_X(-\tilde{D})$ and $\mathcal{L}$ on X . We put

$$\mathcal{M}_X \cong \mathcal{O}_X(-\tilde{D}) \otimes (\mathcal{L}^{-1})^{\otimes n}$$

and obtain for its restriction $\mathcal{M}_{X_K}$ to X_K

$$\mathcal{M}_{X_K} = \mathcal{O}_{X_K}(-D) \otimes (\mathcal{L}_{X_K}^{-1})^{\otimes n} \cong \mathcal{O}_{X_K} \quad .$$

Hence the invertible sheaf $\mathcal{M}_X$ is trivial on the generic fibre and can therefore be written as

$$\mathcal{M}_X = p^*(\mathcal{F})$$

for some $\mathcal{F} \in \operatorname{Pic} U$. We make now the base extension $U' \longrightarrow U$ and obtain the curve $p':X' \longrightarrow U'$ and the invertible sheaves $\mathcal{M}_{X'}, \mathcal{L}', \mathcal{O}_{X'}(-\tilde{D}')$. Since K' is the Hilbert class field of K the sheaf $\mathcal{F} \in \operatorname{Pic} U$ becomes trival

in Pic U' . Denote the resulting sheaf in Pic U' by F'.
Then

$$M_{X'} \cong p_*(F') \cong p_*(O_{U'}) \cong O_{X'}$$

and we have proved the following result.

Proposition 4.7. *Let* X, K, X_K, D *be as above and suppose that* $O_{X_K}(-D) \cong L_{X_K}^{\otimes n}$ *for some invertible sheaf* L_{X_K} *on* X_K *and some integer* $n \geq 1$ *. Then there exists an abelian unramified extension* $K' \supseteq K$ *(the Hilbert class field) of finite degree with ring of integers* R' *and* $U' = U \times_{\operatorname{spec} R} \operatorname{spec} R'$, *a divisor* $\tilde{D}$ *flat over* U *on* $X' = X \times_U U'$ *, and an invertible sheaf* L' *on* X *such that*

$$O_{X'}(-\tilde{D}') \cong L'^{\otimes n} .$$

These sheaves are obtained by base extension $U' \longrightarrow U$ *from the sheaves* $O_{X_K}(-D)$ *,* L_{X_K} *extended to all of* U' *.*

Next consider the situation of Proposition 4.7 and define the $O_{X'}$ -algebra A' on X' by putting

$$L'^{-i} = (L'^{-1})^{\otimes i} \qquad (0 \leq i \leq n)$$

and

$$A = O_{X'} \oplus L'^{-1} \oplus \ldots \oplus L'^{-(n-1)} ,$$

and defining the multiplication on A as

$$L'^{-i}\, L'^{-j} \longrightarrow \begin{cases} L'^{-(i+j)} & \text{if } i+j<n \\ L'^{-(i+j-n)} & \text{otherwise} \end{cases}$$

for $0 \leq i,j \leq n-1$ using the isomorphism

$$\mathcal{O}_{X'}(-\tilde{D}') \cong L'^{\otimes n}$$

which gives us a homomorphism

$$L'^{-n} \longrightarrow \mathcal{O}_{X'} \quad .$$

By Proposition 4.5 we get then a curve *spec* A' over U'. This curve is a ramified covering of X'. Denote it by Y'. Then Y' is a curve over U' which is smooth at the places where $\tilde{D}'$ is smooth and n invertible. This can be easily verified by local considerations. So if the generic fibre $Y'_{K'}$ of Y' is smooth the curve $Y'_{K'}$ has good reduction outside a fixed set of places which depends only on the set of bad places of X, the divisor D and n.

4.4 *The construction of Kodaira-Parshin*

The main step in the deduction of the Mordell conjecture consists of the construction of Kodaira-Parshin (see [Pa]). For this fix a number field K, a finite set of finite places of K *that contains all places* v *of* K *with* $v|2$ and a smooth curve X defined over K with good reduction outside of S. A rational point

$$P : \operatorname{spec} K \longrightarrow X$$

determines an embedding of X into the Jacobian $J(X)$ of X. Henceforth we assume that the genus g of X is at least two. The Jacobian $J(X)$ is also defined over K and the embedding $X \longrightarrow J(X)$ is given by sending a point Q of X to the sheaf $\mathcal{O}_X(Q - P)$. The Jacobian $J(X)$ has good reduction outside of S. Consider now the unramified covering $X^{(2)} \longrightarrow X$ induced by the multiplication by 2 on $J(X)$, i.e. defined by the commutative diagram

$$\begin{array}{ccc} X^{(2)} & \longrightarrow & J(X) \\ \downarrow & & \downarrow 2 \\ X & \longrightarrow & J(X) \end{array}$$

such that $X^{(2)}$ is the pull-back. Then the curve $X^{(2)}$ is defined over K and has good reduction outside of S (note that by definition $v \in S$ for $v|2$). Its genus $g(X^{(2)})$ can be easily determined: First the degree of the covering is equal to 2^{2g}. By Hurwitz (see [Ha]) we get

$$g^{(2)} := g(X^{(2)}) = 2^{2g}(g - 1) + 1 .$$

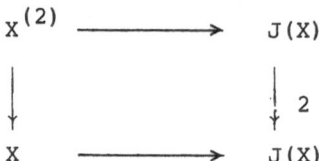

Let D be the inverse image of the divisor P on X. This is a divisor of degree 2^{2g} which is rational over K . Note that it depends on P . In order to apply the results of the last section we need a simple Lemma.

<u>Lemma 4.8.</u> *There exists a finite extension* $K^{(2)} \supseteq K$ *not depending on* $P \in \mathbf{X}(K)$ *such that* $K^{(2)}$ *is unramified outside of* S *with the following property: Let for* $P \in \mathbf{X}(K)$ *be* D *constructed as above. Then there exists an effective divisor* D' *on* $X^{(2)}$ *defined over* $K^{(2)}$ *such that* D *is linearly equivalent to* $2D'$.

Proof. For a given $P \in X(K)$ let K_P be the smallest field containing K such that each point in the fiber over P of the covering $X^{(2)} \longrightarrow X$ becomes K_P-rational. This is an extension of degree at most equal to 2^{2g} and unramified outside of S (note that the places dividing 2 are in S). Apply now Hermite-Minkowski (Theorem 2.5) to obtain $K^{(2)}$.

In order to obtain D' we proceed as follows. The support $|D|$ of D in the Jacobian $J(X)$ is isomorphic to the group $(\mathbb{Z}/2\mathbb{Z})^{2g}$. Find subsets γ' and γ'' of the latter such that

$$\gamma' \cap \gamma'' = \emptyset \ ,$$
$$\gamma' \cup \gamma'' = (\mathbb{Z}/2\mathbb{Z})^{2g} \ ,$$
$$\# \gamma' = \# \gamma'' \ ,$$
$$\sum_{x' \in \gamma'} x' = \sum_{x'' \in \gamma''} x'' \ .$$

This decomposition induces a decomposition

$$D = D' + D''$$

where D' corresponds to γ' and D'' to γ''.

The last property in the definition of γ' and γ'' implies that

$$D' - D'' \sim 0$$

or equivalently

$$D' \sim D'' .$$

Hence

$$D \sim 2D'$$

as desired. Obviously the divisor D' is $K^{(2)}$-rational. This proves the Lemma.

We make now the following base change:

$$\begin{array}{ccccc} \operatorname{spec} K' & \longrightarrow & \operatorname{spec} K^{(2)} & \longrightarrow & \operatorname{spec} K \\ \big\lfloor & & \xrightarrow{\qquad\qquad\qquad\qquad} & & \big\uparrow \end{array}$$

where $\operatorname{spec} K^{(2)} \longrightarrow \operatorname{spec} K$ is defined by Lemma 4.8 and $\operatorname{spec} K' \longrightarrow \operatorname{spec} K^{(2)}$ is defined by Proposition 4.7. We shall now replace everything by its corresponding object after this base change and in order to simplify the notations we still write for them $P, D, D', X, J(X), X^{(2)}$ etc.

They are schemes over spec R'. Since by Lemma 4.8

$$\mathcal{O}_{X^{(2)}}(-D) \cong \mathcal{L}^2$$

for

$$\mathcal{L} = \mathcal{O}_{X^{(2)}}(-D')$$

we can apply the techniques of section 4,3 to obtain a curve

$$Y = Y_P$$

over U' which is a covering of degree 2 of $X^{(2)}$ and ramifies exactly at D . Furthermore it has bad reduction at most at those places where $X^{(2)}$ has bad reduction and those dividing 2. Hence it is a covering

$$Y_P \xrightarrow{f_P} X$$

of X of degree 2^{2g+1} which ramifies only at P . We have therefore proved the following result.

<u>Proposition 4.9.</u> *Let* K *be a number field,* R *its ring of integers,* S *a finite set of finite places containing all places* v *with* $v|2$, $U = \operatorname{spec} R - S$ *and* $X \to U$ *a curve over* U *of genus* $g \geq 2$. *Then there exists a finite extension* $K' \supseteq K$ *with the following property: If* R' *is the ring of integers of* K', S' *the set of places lying over* S *and* U' = spec R'−S' *then for each rational point*

$P \in X(K)$ *there exists a curve* Y_P *over* U' *such that the generic fibre* $Y_{P,K'}$ *of* Y_P *is a covering of* $X_{K'} = X \otimes K'$ *of degree* 2^{2g+1} *that is ramified exactly at* P. *The genus of* Y_P *is equal to* $2^{2g-1}(4g - 3) + 1$.

4.5 *Mordell's conjecture*

We are now able to prove the following result conjectured by Mordell. Let K be as usual a number field.

Theorem 4.10. *Let* X/K *be a smooth curve of genus* $g \geqq 2$. *Then* $X(K)$ *is finite.*

Remarks. 1. Let S be the set of places of K at which X has bad reduction together with the places which divide 2 or 3.

2. Without loss of generality we may assume that the 12-division points in the Jacobian are K-rational.

Proof of Theorem 4.10. By Theorem 4.4 the set of curves $Y_{P,K}$ constructed in Proposition 4.9 is finite up to isomorphism. It remains to show that there are only finitely many coverings

$$\begin{array}{c} Y \\ \downarrow f \\ X \end{array}$$

which are ramified exactly at a fixed point P of given degree and fixed genus $g(Y)$ of Y. But this follows from the fact that there are only finitely many dominant morphisms $f : Y \longrightarrow X$ if the genus X is at least equal to two.

§5 Siegel's Theorem on integer points

Suppose that, as usual, K is a number field, S a finite set of finite places and R_S the ring of S-integers of K. Let X/K be a smooth curve and D an ample divisor on X. Then let $Y = X - |D|$ and $Y(R_S)$ be the set of S-integer points on Y. Then Siegel proved for $S = \emptyset$ the following result which was extended later on by Mahler to arbitrary S.

Theorem 5.1. *Suppose that* $Y(R_S)$ *is infinite. Then the genus of* X *is equal to zero and* Y *is isomorphic to* $\mathbb{G}_a$ *, the additive group, or* $\mathbb{G}_m$ *, the multiplicative group.*

We shall give now a proof of this result using only Mordell's conjecture. We consider first the case that the genus g of X is zero and $|D|$ consists of at least and then without loss of generality exactly 3 different points. Then

$$Y \cong \mathbb{P}^1 \setminus \{0,1,\infty\} .$$

Let $U = \mathbb{P}^1 \setminus \{o,1,\infty\}$ and

$$\begin{array}{c} X' \\ \downarrow P \\ \mathbb{P}^1 \end{array}$$

be a covering of degree 3 fully ramified at $0,1,\infty$. Then the genus X' can be calculated and one obtains by Hurwitz

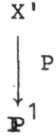

$$2g(X') - 2 = 3(g(X) - 2) + 6 \quad .$$

Hence

$$g(X') = 1 \quad .$$

Let $V = p^{-1}(U)$. Then $V \xrightarrow{p} U$ is proper and étale and an integer point σ: spec $R_S \longrightarrow U$ lifts to a point σ : spec $R_S \longrightarrow V$ over a finite extension K_σ of K unramified outside a set of places T independent of σ (T contains the places of bad reduction of X and the places where D has bad reduction).

The degree of K_σ over K is at most equal to 3. So we find a finite extension $K' \supseteq K$ that contains all the fields K_σ for $\sigma \in Y(R_S)$ (by Theorem 2.5). Hence we may assume that $K' = K$. It remains to show that $X'(R_S)$ is finite. But V is an elliptic curve with 3 points missing. Therefore it is sufficient to show that on an elliptic curve E with one point P missing the set of S-integral points is finite. Let $P = 0$, the point at infinity of E, and $E' = E \setminus 0$. Then as in the last section one constructs a smooth curve X over K which is a covering of E of degree 8 and which ramifies at 0 and which has genus $g(X) = 3$ (see Proposition 4.9). Again an integer point

$$\sigma\text{: spec } R_S \longrightarrow E'$$

lifts to a point

$$\sigma: \operatorname{spec} R_S \longrightarrow X$$

after an eventually finite extension of K as before. Since $g(X') \geq 2$ we may apply Theorem 4.10 and find that $E'(R_S)$ is finite.

References

[De] P. Deligne, Preuve des conjectures de Tate et Shafarevitch [d'après G. Faltings] , Sém. Bourbaki, Exposé 616,1983.

[EGA IV] A. Grothendieck, Elements de Géométrie Algébrique IV (seconde partie), Publ. Math. IHES 24 (1965).

[Fa] G. Faltings, Heights, this volume.

[Gru] F. Grunewald, Some facts from the theory of group schemes, this volume.

[Ha] R. Hartshorne, Algebraic Geometry, Springer Verlag (1977).

[Ma] J. Martinet, Character Theory and Artin L-functions; in: Algebraic Number fields (A. Fröhlich, ed.), Proc. LMS Symp. Durham, Acad. Press (1977).

[Mu] D. Mumford, J. Fogarty, Geometric Invariant Theory, 2nd edition, Springer Verlag, 1982.

[Pa] A.N. Parshin, Algebraic curves over function fields I, Math. USSR Jzvestija 2, 1145-1170 (1968).

[Rei] I. Reiner, Maximal Orders, Academic Press, London-New York-San Francisco (1975).

[Sc] Séminaire Cartan, 1960/61.

[Sch] N. Schappacher, Tate's conjecture on the endomorphisms of abelian varieties, this volume.

[Se 1] J.P. Serre, Abelian ℓ-adic representations and elliptic curves, W.A. Benjamin, New York, Amsterdam (1968).

[Se 2] J.P. Serre, Quelques applications du Théorème de densité de Chebotarev, Publ. Math. IHES 54 (1981).

[SGA 7I] A. Grothendieck, Groupes de Monodromie en Géométrie Algébrique, SLN 288.

[Sz] L. Szpiro, La conjecture de Mordell [d'après G. Faltings], Sem. Bourbaki, Exposé 619, 1983.

[We] A. Weil, Adeles and Algebraic Groups, Prog. Math.23.

VI

COMPLEMENTS TO MORDELL

Gerd Faltings

Contents:

§ 1 Introduction

The purpose of this chapter is to give some additional results, mainly about generalizations to finitely generated extensions of $\mathbb{Q}$. Similar results have been obtained by other people, and on occasion I have used their arguments instead of my original ones. More precisely, we obtain the following facts:

Choose a finitely generated extension field K of $\mathbb{Q}$ and let $R \subseteq K$ denote a finitely generated smooth $\mathbb{Z}$-algebra, with field of quotients K . As before, $\pi = \mathrm{Gal}(\overline{K}/K)$ is the absolute Galois-group of K .

For an abelian variety A over K , π acts continuously on the Tate-module $T_l(A)$ (l a prime). We have:

<u>Theorem 1</u> (Tate-Conjecture)

a) $T_l(A) \otimes_{\mathbb{Z}_l} \mathbb{Q}_l$ is a semisimple π-module

b) The map

$\mathrm{End}_K(A) \otimes_{\mathbb{Z}} \mathbb{Z}_l \to \mathrm{End}_\pi(T_l(A))$

is an isomorphism

c) Except for finitely many primes l , the image of the mapping

$\mathbb{Z}_l[\pi] \to \mathrm{End}_{\mathbb{Z}_l}(T_l(A))$

is the full commutator of $\mathrm{End}_K(A)$

Theorm 2: (Shafarevich-conjecture)
Up to isomorphism, there exist only finitely many abelian varieties of a given dimension g over K, which have good reduction at all primes $\underline{p} \subset R$ of height one. The same holds if we consider d-fold polarized abelian varieties, for some integer $d > 0$.

Theorem 3: (Mordell-conjecture)
Any curve over K of genus bigger than one has only finitely many rational points.

Theorem 4:
If A is an abelian variety over a field L of characteristic zero, and $X \subset A$ a curve of genus bigger than one, then for any finitely generated subgroup $\Gamma \subset A(L)$, $\Gamma \cap X$ is finite.

Theorem 5:
The mapping

$$\mathrm{End}_K(A) \to \mathrm{End}_\pi(A(K)) \quad \text{is}$$

is an isomorphism
We also describe some ideas of A.N. Parshin and J.G. Zarhin, which give an effective bound for the number of rational points on a curve of genus bigger than one.

Most results are proven by reduction to the case of number-fields. This is achieved via complex Hodge-theory. In the next paragraph we give the necessary preliminaries.

§ 2 Preliminaries

1.) The Čebotarev-density theorem

Let $S=\mathrm{Spec}(R)$ with $R\subseteq K$ as before, R smooth over $\mathbb{Z}$. Let $\pi_1(S)$ be the étale fundamental group of S. (with respect to some geometric point). If $x\in S$ is a closed point, its residue field $k(x)$ is finite with $N(x)$ elements, and we obtain a conjugacy class F_x in $\pi_1(S)$ containing the canonical generator of $\mathrm{Gal}(\overline{k(x)}/k(x))\cong\hat{\mathbb{Z}}$. By abuse of notation we will often speak just of the element $F_x\in\pi_1(S)$, which is determined up to conjugation.

Theorem: (Čebotarev)

The conjugacy classes of the F_x are dense in $\pi_1(S)$.

sketch of proof: We may replace S by an open subscheme, hence assume that a fixed prime l is invertible in R. We have to show that for any continuous surjection of $\pi_1(S)$ onto a finite group G the images of the F_x meet any conjugacy class of G. Following the proof in the numberfield case it suffices it for any irreducible representation χ on G over a finite extension E of $\mathbb{Q}$, the L-series

$$L(s,\chi) = \prod_x \det(1-N(x)^{-s}\cdot\chi(F_x))^{-1}$$

is holomorphic for $\mathrm{Re}(s) > d = \dim(S)$, can be continued meromorphically to $\mathrm{Re}(s) > d-\frac{1}{2}$, and has at $s=d$ either a pole of first order (if χ = trivial representation), or neither a pole nor a zero (if $\chi\neq$ trivial).

Using Brauer's induction theorem one reduces to abelian characters

$$\chi : \pi \to \mu = \text{roots of unity.}$$

By Grothendieck's formula, if F denotes the étale l-adic sheaf associated to χ:

$$L(s,\chi) = \prod_{i=0}^{2(d-1)} \left(\prod_p \det(1-p^{-s}\cdot F_p | H_c^i(S \otimes \bar{\mathbb{F}}_p, F)) \right)^{(-1)^{i+1}}$$

It is known that the factors for $i < 2(d-1)$ are holomorphic and non-zero for $\mathrm{Re}(s) > d - \frac{1}{2}$, and that $H_c^{2(d-1)}(S \otimes \bar{\mathbb{F}}_p, F)$ is dual to $H^0(S \otimes \bar{\mathbb{F}}_p, F^*)(d-1)$, hence we have to worry only about

$$\prod_p \det(1-p^{d-1-s} F_p^{-1} H^0(S \otimes \bar{\mathbb{F}}_p, \check{F}))^{-1}$$

This is essentially the L-series for the representation of $\mathrm{Gal}(\bar{\mathbb{Q}}/\mathbb{Q})$ on the dual of $H^0(S \otimes \bar{\mathbb{Q}}, \check{F})$, with a shift $d-1$ in the variable s. If L denotes the algebraic closure of $\mathbb{Q}$ in K, this representation is induced from the $\mathrm{Gal}(\bar{L}/L)$ representation on $H^0(S \otimes \bar{L}, \check{F})$. But $H^0(S \otimes \bar{L}, \check{F})$ vanishes, unless $\check{F}$ is trivial on $S \otimes \bar{L}$, that is, unless χ is given by a character of $\mathrm{Gal}(\bar{L}/L)$. In this case we have to consider the L-series of this character, and its behaviour is known.

2.) Decomposition groups

Suppose X is an geometrically irreducible normal algebraic variety over a numberfield L, of dimension at least one. The fundamental group $\pi_1(X)$ of X is then an extension of the geometric fundamental group $\pi_1^0(X) = \pi_1(X \otimes_L \bar{L})$ by the Galoisgroup $\mathrm{Gal}(\bar{L}/L)$:

$$0 \to \pi_1{}^{o}(X) \to \pi_1(X) \to \mathrm{Gal}(L/\bar{L}) \to 0$$

To any $\bar{L}$-valued point $p \in X(\bar{L})$ of X corresponds a decomposition group

$$D_p \subseteq \pi_1(X) .$$

The mapping from D_p to $\mathrm{Gal}(\bar{L}/L)$ is an injection, and gives an isomorphism of D_p with some Galoisgroup $\mathrm{Gal}(\bar{L}/L_1)$, where $L_1 \subseteq \bar{L}$ is the field of definition of p . Hence the semidirect product $\pi_1{}^{o}(X) \rtimes D_p$ has finite index in $\pi_1(X)$.

3.) Complex Hodge-Theory

Consider a smooth geometrically irreducible algebraic variety X over a number-field L , similar as in 2.) . If we choose an embedding $L \hookrightarrow \mathbb{C}$, $\pi_1{}^{o}(X)$ is the profinite completion of the topological fundamental group $\pi_1(X(\mathbb{C}))$. This gives us valuable information, for example that it is finitely generated.

Furthermore, if

$$\phi : A \to X$$

is an abelian variety over X , and $p \in X(\bar{L}) \subseteq X(\mathbb{C})$ a geometric point, the action of $\pi_1{}^{o}(X)$ on $T_l(A)$ (l a prime) is induced from the representation of $\pi_1(X(\mathbb{C}))$ on $H_1(A(p),\mathbb{Z}) = T(A)$.

This representation has the following wellknown properties: (Déligne, Hodge II)

a) $T(A) \otimes_{\mathbb{Z}} \mathbb{Q}$ is a semisimple $\pi_1(S(\mathbb{C}))$-module.

b) Consider the injection

$$\mathrm{End}_{X \otimes_L \mathbb{C}}(A) \hookrightarrow \mathrm{End}_{\pi_1(S(\mathbb{C}))}(T(A)) :$$

An endomorphism of $T(A)$ commuting with $\pi_1(S(C))$ is already in the image if it induces an endomorphism of one fibre of ϕ, for example the fibre at p.

Thus:

$$\mathrm{End}_{X \otimes_L \mathbb{C}}(A) = \mathrm{End}_{\pi_1(S(\mathbb{C}))}(T(A)) \cap \mathrm{End}_{\mathbb{C}}(A(p) \otimes \mathbb{C})$$

4.) Hermite-Minkowski

Let $S = \mathrm{Spec}(R)$ be as in 1.), R smooth and finitely generated over $\mathbb{Z}$.

Theorem: (Hermite-Minkowski)

Suppose G is a finite group. Then there exist only finitely many continuous homomorphism

$$\rho : \pi_1(S) \to G$$

<u>sketch of proof</u>:

Let L be the algebraic closure of $\mathbb{Q}$ in K (K = quotient-field of R). Then

$$X = S \otimes_{\mathbb{Z}} \mathbb{Q}$$

is geometrically irreducible over L , and $\pi_1(X)$ surjects onto $\pi_1(S)$.

Choose a geometric point $P \in X(\bar{L})$. Then $\pi_1{}^{o}(X) \rtimes D_P$ has finite index in $\pi_1(X)$, so that it suffices to show that the various ρ's restrict to finitely many morphisms from $\pi_1{}^{o}(X) \rtimes D_P$ to G . But their restrictions to D_P give only finitely many different elements by the classical Hermite-Minkowski-theorem, and the same is true for the restricions to $\pi_1{}^{o}(X)$,because this group is topologically finitely generated.

§ 3 THE TATE-CONJECTURE

Theorem 1:

Suppose that K is a finitely generated extension of $\mathbb{Q}$, A an abelian variety over K, $T_l(A)$ its Tate-Module (for some prime l),

$$\rho_l : \pi = \mathrm{Gal}(\bar{K}/K) \to \mathrm{Aut}(T_l(A))$$

the corresponding representation.

Then

a) $T_l(A) \otimes_{\mathbb{Z}_l} \mathbb{Q}_l$ is a semisimple π-module

b) $\mathrm{End}_K(A) \otimes_{\mathbb{Z}} \mathbb{Z}_l \cong \mathrm{End}_\pi(T_l(A))$

c) For almost all l, the subalgebra of $\mathrm{End}_{\mathbb{Z}_l}(T_l(A))$ generated by $\rho_l(\pi)$ is the full commutator of $\mathrm{End}_K(A)$.

Corollary:

Up to isomorphy, there exist only finitely many abelian varieties B over K which are isogeneous to A.

Proof:

We start by some general remarks. Properties a) and b) imply that for any prime l the subalgebra of $\mathrm{End}_{\mathbb{Z}_l}(T_l(A))$ generated by π has finite index in the commutator of $\mathrm{End}_K(A)$. To prove c), we may restrict ourselves to primes l for which $\mathrm{End}_K(A) \otimes_{\mathbb{Z}} \mathbb{Z}/_{l\mathbb{Z}}$ is a semisimple $\mathbb{Z}/_{l\mathbb{Z}}$-algebra. For those l, property c) holds if and only if $T_l(A)/l \cdot T_l(A)$ is a semisimple π-module, whose π-endomorphisms are given by $\mathrm{End}_K(A) \otimes_{\mathbb{Z}} \mathbb{Z}/_{l\mathbb{Z}}$. If $\pi' \subseteq \pi$ is a closed subgroup with $[\pi : \pi']$ finite and prime to l, it suffices to show this property for π' instead of π . This also applies to a) and b), and we thus may assume the following hypotheses:

Let $L \subseteq K$ denote the algebraic closure of $\mathbb{Q}$ in K . Then there exists a smooth, geometrically irreducible scheme X over L , with function field K , such that A extends to an abelian variety over A . Furthermore, X has a rational point $p \in X(L)$.

Thus π acts on $T_l(A)$ via its quotient $\pi_1(X)$. If we choose an embedding $L \hookrightarrow \mathbb{C}$, $\pi_1(X)$ decomposes as a semidirect product

$$\pi_1(X) = \pi_1^0(X) \rtimes D_p \quad ,$$

where π_1^0 X) is the profinite completion of the topological fundamental group $\pi_1(X(\mathbb{C}))$.

If $A(p)$ denotes the fibre of A over p , properties a) , b) and c) are known for $A(p)$ (with the action of $D_p \cong \mathrm{Gal}(\bar{L}/L)$). We let $T(A) = H_1(A(p)(\mathbb{C}), \mathbb{Z})$, so that $T_l(A) = T(A) \otimes_{\mathbb{Z}} \mathbb{Z}_l$, and the action of π_1^0 on $T_l(A)$ is derived from the action of $\pi_1(X)(\mathbb{C}))$ on $T(A)$. The rest is easy:

a) $T_l(A) \otimes_{\mathbb{Z}_l} \mathbb{Q}_l$ is a semisimple π-module: let $\mathfrak{g}, \mathfrak{g}^0$ and $\mathfrak{f}$ denote the Lie-algebras of the compact l-adic groups $\rho_l(\pi)$, $\rho_l(\pi^0)$ and $\rho_l(D_p)$. We have to show that $\mathfrak{g}$ is reductive in $T_l(A) \otimes_{\mathbb{Z}_l} \mathbb{Q}_l$. We know that this already holds for $\mathfrak{g}^0$ (by complex Hodge-theory) and $\mathfrak{f}$ (Tate-conjecture for

$A(p))$. But $\mathfrak{g}^{\circ}$ is an ideal in $\mathfrak{g}$, and $\mathfrak{g} = \mathfrak{g}^{\circ} + \mathfrak{f}$.
The claim follows.

b) $\mathrm{End}_K(A) \otimes_{\mathbb{Z}} \mathbb{Z}_l \xrightarrow{\sim} \mathrm{End}_{\pi}(T_l(A))$: We have an injection of left into right. Furthermore, we know that $\mathrm{End}_K(A) = \mathrm{End}_X(A)$

$$= \mathrm{End}_{X \otimes_L \mathbb{C}}(A) \cap \mathrm{End}_L(A(p))$$

$$= \mathrm{End}_{\pi_1(X(\mathbb{C}))}(T(A)) \cap \mathrm{End}_L(A(p))$$

Tensoring with $\mathbb{Z}_l$ and applying the Tate-conjecture to $A(p)$ gives:

$$\mathrm{End}_K(A) \otimes_{\mathbb{Z}} \mathbb{Z}_l = \mathrm{End}_{\pi_1^{\circ}}(T_l(A)) \cap \mathrm{End}_{D_p}(T_l(A)$$

$$= \mathrm{End}_{\pi}(T_l(A)) .$$

c) For almost all l , $\rho_l(\pi)$ generates the full commutator of $\mathrm{End}_{\nu}(A)$:

Taking into account a) and b) we have to show that there exists a subalgebra $M \subseteq \mathrm{End}_{\mathbb{Z}}(T(A))$ (of finite index in the commutator of $\mathrm{End}_K(A) = \mathrm{End}_X(A)$) , such that for all l $M \otimes_{\mathbb{Z}} \mathbb{Z}_l$ is the subalgebra generated by $\rho_l(\pi)$.

If we replace π_1 by π_1° , such an algebra is given by the image of $\mathbb{Z}[\pi_1(X(\mathbb{C}))]$. The same can be said about $D_p \subseteq \pi_1(X)$, by using the case of number-fields. We take for M the algebra generated by those two subalgebras. The corollary follows, because $M \otimes_{\mathbb{Z}} \mathbb{Q}$ is semisimple, and abelian varieties B isogeneous to A correspond to M-lattices in $T(A) \otimes_{\mathbb{Z}} \mathbb{Q}$. By the Jordan-Zassenhaus-theorem, there are only finitely many isomorphism classes of such lattices.

§ 4 The Shafarevich-Conjecture

Theorem 2: Let S be an integral scheme, smooth and of infinite type over $\mathbb{Z}$. For any g, there exist up to isomorphism only finitely many abelian varieties A of dimension g over the function-field K of S, which extend to abelian varieties over some open set $U \subset S$ with codim $(S-U) \geq 2$.
The same holds for isomorphism classes of d-fold polarized abelian varieties, for any integer d.

Proof: The two statements are equivalent, so we only show the first one. The corollary to the Tate-conjecture (Th.1) implies that it suffices to prove finiteness up to isogeny, and by the Tate-conjecture we only need to consider the isomorphism-classes of the Galois-representations $T_l(A) \otimes_{\mathbb{Z}_l} \mathbb{Q}_l$.
We may assume that l is invertible on S. If A extends to an abelian variety over S, we know that $\pi_1(S)$ acts semi-simple on $T_l(A) \otimes_{\mathbb{Z}_l} \mathbb{Q}_l$, and that this representation is pure of weight $1/2$ (that is, for $x \in S$ a closed point, the Frobenius F_x has eigenvalues of absolut value $N(x)^{1/2}$).
We show that these properties also hold if A has only good reduction up to codimension 2: By purity of the branch-locus, the representation of $\mathrm{Gal}(\bar{K}/K)$ on $T_l(A)$ factors over its quotient $\pi_1(S)$. This representation is also pure of weight $1/2$, because for any closed point $x \in S$ we can find a proper birational morphism $\varphi: \tilde{S} \to S$, such that $\tilde{S}$ is regular and $\varphi^{-1}(x)$ is a divisor in $\tilde{S}$ (take the blow-up in x, for example). As $\varphi^*(T_l(A))$ is unramified on $\tilde{S}$, $\varphi^*(A)$

extends to an abelian variety over some open set $\tilde{U} \subset \tilde{S}$, whose complement has codimension at least two. Thus there exists a closed point $y \in \tilde{U} \cap \varphi^{-1}(x)$, and the eigenvalues of F_y have the correct absolute value. As F_y is a power of F_x, we are done.

Now the original proof of the finiteness of isogeny classes applies (Ch.V, Th. 2.8), since we only need

a) Hermite-Minkowski

b) Čebotarev

c) The Tate-conjecture.

Thus the proof of theorem 2 is complete.

By the Parshin-construction, we obtain the Mordell-conjecture

Theorem 3:

Let X be a curve of genus $g \geq 2$, defined over a finitely generated extension K of $\mathbb{Q}$. Then $X(K)$ is finite.

Remark:

Another way to show this is to make use of the Mordell-conjecture for function-fields (Manin, Grauert) and reduce to number-fields.

The Mordell-conjecture is equivalent to the following old conjecture:

Theorem 4:

Let L be a field of characteristic zero, A on abelian variety over L , and $X \subset A$ a curve of genus bigger than one. If $\Gamma \subset A(L)$ is a finitely generated abelian group, $\Gamma \subset X(L)$ is finite.

proof:

There exists a finitely generated extension of $\mathbb{Q}$ contained in L, $K \subseteq L$, such that A and X are defined over K, and $\Gamma \subset A(K)$. Then $\Gamma \cap X(L) \subseteq X(K)$, and this is finite.

Remark:

By results of M. Raynaud, this also holds if we assume that Γ has only finite rank.

§ 5 Endomorphisms

Again K is a finitely generated extension of $\mathbb{Q}$, $\bar{K}$ its algebraic closure, $\pi = \mathrm{Gal}(\bar{K}/K)$. π operates continuously on the divisible group $A(\bar{K})$, and we have an exact sequence

$$0 \to A(\bar{K})_{tors} \to A(\bar{K}) \to A(\bar{K})_{ntors} \to 0$$

with

$$A(\bar{K})_{tors} = \bigoplus_{l} T_l(A) \otimes \mathbb{Q}_l/\mathbb{Z}_l$$
$$A(\bar{K})_{ntors} = A(\bar{K})/A(\bar{K})_{tors}$$

$A(\bar{K})_{ntors}$ is a vectorspace over $\mathbb{Q}$, and it's the union of finite-dimensional π-modules. More precisely, if $\pi' \subseteq \pi$ is a closed subgroup of finite index, the space of π'-invariants in $A(\bar{K})_{ntors}$ is finite-dimensional (by the Mordell-Weil theorem)

There is a natural injection

$$\mathrm{End}_K(A) \to \mathrm{End}_{\pi}(A(\bar{K})),$$

and we want to prove that it is an isomorphism. We proceed by several lemmas.

Lemma 1:

Let M be a π-module which is a subquotient of $A(\bar{K})_{tors}$. Then

$$\mathrm{Hom}_{\pi}(A(\bar{K})_{ntors}, M) = 0$$

proof:

If $\pi' \subseteq \pi$ is a normal subgroup of finite index, we show that

$$\mathrm{Hom}_{\pi}(A(\bar{K})^{\pi'}_{ntors}, M) = 0$$

Choose a subring $R \subseteq K$, smooth over $\mathbb{Z}$ etc. (as always), such that the normalization R' of R in the field $K' = \bar{K}^{\pi'}$ is étale over R, such that A extends to $S = \mathrm{Spec}(R)$, and such that the $A(K')$-valued points of A extend to R'-valued points. We furthermore may assume that M is an l-torsion group, for some prime l, and that l is invertible in R.

Then the π-operation on $A(\bar{K})^{\pi'}_{ntors}$ and M is induced from a $\pi_1(S)$-operation. For this operation $A(\bar{K})^{\pi'}_{ntors}$ is pure of weight zero (each F_x has roots of unity as eigenvalues, because $A(\bar{K})^{\pi'}_{ntors}$ is a finite-dimensional $\mathbb{Q}$-vectorspace), while M is pure of weight $\frac{1}{2}$. (F_x has eigenvalues of absolute value $N(x)^{1/2}$). So there cannot exist a nontrivial $\pi_1(S)$-morphism.

Lemma 2:

$$\mathrm{Hom}_{\pi}(A(\bar{K}) \ , \ A(\bar{K})_{tors}) = 0$$

proof:

By lemma 1, this injects into $\mathrm{End}_{\pi}(A(\bar{K})_{tors})$, and by the Tate-conjecture we may assume that A is simple, hence that

$$D = \mathrm{End}_K(A) \otimes_{\mathbb{Z}} \mathbb{Q}$$

is a skew-field.

Let $c \in \mathrm{Ext}^1_{\pi}(A(\bar{K})_{ntors}, A(\bar{K})_{tors})$ denote the class of the extension

$$0 \to A(\bar{K})_{tors} \to A(\bar{K}) \to A(\bar{K})_{ntors} \to 0$$

From the usual proof of the Mordell-Weil theorem one knows that for any prime l and any finite extension $K' = \bar{K}^{\pi'}$ of K , the cup-product with c gives an injection

$$A(K') \otimes \mathbb{Q}_l/\mathbb{Z}_l \hookrightarrow H^1(\pi'; T_l(A) \otimes \mathbb{Q}_l/\mathbb{Z}_l) .$$

Now suppose that $\mathrm{Hom}_{\pi}(A(\bar{K}), A(\bar{K})_{tors}) \neq 0$. Choose a non-zero element ψ in this group.

By lemma 1, $\psi(A(\bar{K})) = \psi(A(\bar{K})_{tors})$, hence

$$A(\bar{K}) = A(\bar{K})_{tors} + \mathrm{Ker}(\psi) ,$$

hence c goes to zero under the mapping

$$\mathrm{Ext}^1_\pi(A(\bar{K})_{ntors}, A(\bar{K})_{tors}) \to \mathrm{Ext}^1_\pi(A(\bar{K})_{ntors}, A(\bar{K})_{tors}/\mathrm{Ker}\psi)$$

Thus for some prime l there exists a π-invariant sublattice $W \subsetneqq T_l(A)$, such that c goes to zero in

$$\mathrm{Ext}^1_\pi(A(\bar{K})_{ntors}, (T_l(A)/W) \otimes_{\mathbb{Z}_l} \mathbb{Q}_l/\mathbb{Z}_l)$$

We show that this cannot happen:

$W \otimes_{\mathbb{Z}_l} \mathbb{Q}_l$ is a subspace of $T_l(A) \otimes_{\mathbb{Z}_l} \mathbb{Q}_l$, invariant under π. By the Tate-conjecture, it thus must be the image of an idempotent e of $D \otimes_{\mathbb{Q}} \mathbb{Q}_l$. There is a natural number n with $n \cdot e \in \mathcal{O} \otimes_{\mathbb{Z}} \mathbb{Z}_l$, and $n(1-e)$ annihilates the image of c in

$$\mathrm{Ext}^1_\pi(A(\bar{K})_{ntors}, T_l(A) \otimes_{\mathbb{Z}_l} \mathbb{Q}_l/\mathbb{Z}_l)$$

and hence also

$$A(K') \otimes \mathbb{Q}_l/\mathbb{Z}_l ,$$

for each finite extension $K' \supseteq K$. If we choose K' in such a way that $A(K')$ contains a non-torsion element, $A(K')$ contains as a submodule of finite index a free $\mathcal{O}$-module of positive rank. Thus $A(K') \otimes \mathbb{Q}_l/\mathbb{Z}_l$ can be annihilated by $n(1-e)$ only if $e=1$, hence $W=T_l(A)$. This is a contradiction.

<u>end of proof of theorem 4:</u>

We have a diagram

$$\begin{array}{ccccc} \mathrm{End}_K(A) & \to & \mathrm{End}_\pi(A(\bar{K})) & \hookrightarrow & \mathrm{End}_\pi(A(\bar{K})_{ntors}) \\ & & \downarrow & & \\ & & \mathrm{End}_\pi(A(\bar{K})_{tors}) & = & \mathrm{End}_K(A) \otimes_{\mathbb{Z}} \hat{\mathbb{Z}} \end{array}$$

It suffices if

$$\mathrm{End}_K(A) \otimes_{\mathbb{Z}} \mathbb{Z}_l \to \mathrm{End}_\pi(A(\bar{K})) \otimes_{\mathbb{Z}} \mathbb{Z}_l$$

is an isomorphism for each l , and we are ready if we show that the mapping

$$\mathrm{End}_\pi(A(\bar{K})) \otimes_{\mathbb{Z}} \mathbb{Z}_l \to \mathrm{End}_\pi(T_l(A))$$

is injective, for each l .
As $\mathrm{End}_\pi(A(\bar{K}))$ is torsion-free, we have to show:

Claim:
If $f_1,\ldots,f_r \in \mathrm{End}_\pi(A(\bar{K}))$ are linearly independant over $\mathbb{Z}$, they are linearly independant over $\mathbb{Z}_l$ as endomorphisms of $T_l(A)$.

proof of claim:
As $\mathrm{End}_\pi(A(\bar{K}))$ injects into $\mathrm{End}_\pi(A(\bar{K})_{\mathrm{ntors}})$, we can find a finite extension K' of K such that the f_i are linearly independant as endomorphisms of $A(K')$. As $\mathrm{End}\,(A(K'))$ us a finitely generated abelian group, there exists a constant d , such that

$$l^n \cdot \mathrm{End}(A(K')) \cap (\mathbb{Z}f_1 + \quad + \mathbb{Z}f_r) \subseteq l^{n-d} \cdot (\mathbb{Z}f_1 + \quad + \mathbb{Z}f_r) ,$$

for $n \geq d$. (Artin-Rees)
If the f_i are not $\mathbb{Z}_l$-independant as endomorphisms of $T_l(A)$, there is a sequence $n_j \in \mathbb{Z}^r$, $n_j=(n_{j1},\ldots,n_{jr})$,

such that

a) not all components of n_j are divisible by l

b) $\sum_{i=1}^{r} n_{ji} f_i \in l^j \operatorname{End}_\pi(T_l(A))$.

Hence $\sum_{i=1}^{r} n_{ji} f_i$ annihilates the l^j-torsion-points of $A(\bar{K})$, so $\sum_{i=1}^{r} n_{ji} f_i \in l^j \cdot \operatorname{End}_\pi(A(\bar{K}))$,

hence

$$\sum_{i=1}^{r} n_{ji} \cdot f_i \in l^j \operatorname{End}(A(K')) ,$$

hence

$n_{ji} \in l^{j-d} \cdot \mathbb{Z}$. For $j > d$, This is a contradiction.

§ 6 EFFECTIVITY

A.N. Parshin and J.G. Zarhin have found a method which leads to an effective bound for the number of rational points on a curve of genus bigger than one, over a numberfield K. We intend to give a sketch.

Let K denote a numberfield, X a curve of genus $g \geq 2$ over K. The Parshin-construction associates to any rational point $x \in X(K)$ an abelian variety $A(x)$, whose dimension is independant of x.

Let us suppose that there exists a rational point $x_0 \in X(K)$. We let $h(x)=h_{\underline{L}}(x)$ denote the height of $x \in X(K)$, measured by the line-bundle $\underline{L}=\mathcal{O}_X(x_0)$. Then $h(x)$ is related to the height of $A(x)$, $h(A(x))$, by

$$h(A(x)) = c_1 \cdot h(x) + O(\sqrt{|h(x)|+1}) ,$$

whith some constant $c_1 > 0$.

We already know that there exist only finitely many isogeny-classes of $A(x)$'s, and we can bound their number if we use the effective Čebotarev-theorem. ([LO])

It is thus sufficient to bound the number of points $x \in X(K)$ for which $A(x)$ is isogeneous to a fixed abelian variety A; If we show that for two such rational points $x_1, x_2 \in X(K)$, the difference in heights $|h(A(x_1))-h(A(x_2))|$ can be bounded

effectively, we may use an old result of Mumford ([M]): The mapping $\lambda: x \mapsto \lambda(x) = \mathcal{O}(x - x_0)$ embeds $X(K)$ into the Mordell-Weill group $J(K)$ †
The Néron-Tate height makes $J(K) \otimes_{\mathbb{Z}} \mathbb{R} = V$ an euclidean vector-space, and for a pair of reals $0 < r,s$ the number of rational points $x \in X(K)$ with $r \leq \|\lambda(x)\| \leq r(1+s)$ is effectively bounded, with the bound depending only on s. As $\|\lambda(x)\|$ is related to $h(x)$ just as $h(A(x))$ by a relation

$$\|\lambda(x)\| = c_2 \cdot h(x) + O(\sqrt{|h(x)|+1}) ,$$

we see that for any $x \in X(K)$ with $A(x)$ isogeneous to A, we have either $\|\lambda(x)\| \leq 1$, or $r \leq |\lambda(x)\| \leq r(1+s)$ with constants r,s independant of x, and such that s can be effectively determined. Thus the number of those x is bounded.

We thus are reduced to bounding the difference of heights in one isogeny-class. So let us consider abelian varieties B isogeneous to a fixed A, and with good reduction outside a given set S of places of K. We may assume that A and all B's are semistable. The Weil-conjectures give an effective number N, such that for any l-isogeny $\varphi: B_1 \to B_2$, with l a prime bigger than N, the heights $h(B_1)$ and $h(B_2)$ are equal.

We are thus reduced to consider l-isogenies for $l \leq N$, or for just one fixed prime l.

† (J=Jacobian of X)

By a series of reduction steps one shows that φ can be factored into a product of finitely many isogenies $\varphi_1,\dots,\varphi_r$, with the number r independant of φ , and each φ_i having one of the following properties:

Either

a) $\deg(\varphi_i) = 1$,

or

b) For each place v of K dividing l , the kernel $G_{i,v}$ of the extension of φ_i to the Néron-models over the local ring O_v is a truncated l-divisible group of some exponent $s \geq 2$. This means that the Tate-module of $G_{i,v}$ is of the form $(\mathbb{Z}/_{l^s\mathbb{Z}})^{h_v}$, and $G_{i,v}$ satisfies the axioms for an l-divisible group "up to order s". Furthermore, the Tate-module of G_i (over K) is of the form $(\mathbb{Z}/_{l^s\mathbb{Z}})^h$. For isogenies φ_i of type a) we know that

$$|h(B_1) - h(B_2)| \leq \frac{1}{2}\log(l) ,$$

so we may assume that $\varphi=\varphi_i$ is of type b) . By a theorem of Grothendieck the truncated l-divisible group $G_v=G_{i,v}$ over the completion $\hat{\mathcal{O}}_v$ may be extended to a full l-divisible group. It thus has invariants d_v and h_v , and we have to show that for s big

$$d = \sum_{v|l} d_v[K_v:\mathbb{Q}_l] = [K:\mathbb{Q}]\cdot h/_2 .$$

The left hand side can be determined by considering the action of $\pi=\mathrm{Gal}(\bar{K}/K)$ and $\tilde{\pi}=\mathrm{Gal}(\bar{\mathbb{Q}})/\mathbb{Q})$ on the Tate-modules. We obtain that the determinant of the action of $\tilde{\pi}$ on the on the induced Tate-module is $\chi_o^d \cdot \varepsilon^h$, where χ_o is the cyclotomic character, and ε the permutation character. Here these characters take values in $(\mathbb{Z}/l^s\mathbb{Z})^*$.

The Weil-conjectures show that either the equality above holds ore l^s divides a certain number $M > 0$ which can be effectively determined. Thus either φ_i does not change heights, or its degree is effectively bounded.

This finishes the argument of Parshin and Zarhin.

BIBLIOGRAPHY:

[L] S.Lang: Division points on curves
Annalli di Matematicce Pura ed Applicata,
ser. 4, 70(1965), 229-234.

[LO] J.C. Lagarias/ A.M. Odlyzko: Effective versions of the Chebotarev density theorem
Proc. Sympos. Univ. Durham 1975, 409-464.
Academic Press, London 1977.

[M] D. Mumford: A remark on Mordell's conjecture
Amer. J. Math. 87(1965), 1007-1016.

[R] M. Raynaud: Courbes sur une varieté abélienne et points de torsion.
Invent. Math. 71(1983), 207-233.

VII

INTERSECTION THEORY ON ARITHMETIC SURFACES

Ulrich Stuhler

Contents:

§ 0 INTRODUCTION

The purpose of this part is to give an introduction to intersection theory on arithmetic surfaces, a theory initiated by S.Yu Arakelov in [A1,2,3] and further developped by G. Faltings in [F]*). The idea, propagated during the last years in particular by L. Szpiro, is roughly to replace or better to enrich algebro-geometric structures at the infinite primes involved by hermitian structures as for example hermitian line bundles, curvatures, volumes etc.

We describe the approach more detailed: Suppose $X \overset{\pi}{\to} B$ is a semistable curve over $B=\mathrm{Spec}(R)$, R the ring of algebraic integers in the field K . Suppose, D_1 and D_2 are divisors (in the usual sense) on X . We want to associate an intersection number $\langle D_1,D_2\rangle$. This is easy if by chance one of the divisors, say D_1 , is vertical with respect to π , $D_1 \subseteq \pi^{-1}(v), v\in B$, and D_1 irreducible. We consider the line bundle $\mathcal{O}_X(D_2)$ on X and obtain

$$\langle D_1,D_2\rangle = \log(q_v)\deg(\mathcal{O}_X(D_2) \mid_{D_1}) ,$$

the degree of the restriction of $\mathcal{O}_X(D_2)$ to D_1 , multiplied with $\log(q_v)$, $q_v=\#k(v)$, the order of the residue field at v . It is this definition which can be made to work in general.

*) See also P. Hriljac [H].

Suppose $D_1 = s(B)$ is a section - this is the critical case. The idea is to put hermitian structures on all the line bundles $\mathcal{O}_X(D_2)$. Then we can consider the hermitian line bundle $s^*(\mathcal{O}_X(D_2))$ on B . We have a degree map for these and can define

$$<D_1, D_2> = \deg s^*(\mathcal{O}_X(D_2))$$

in perfect coincidence with the definition above.

The problem is to find a consistent system of hermitian metrics on the line bundles $\mathcal{O}(D)$ on B . This will be done in §1 and once this is achieved the elementary properties of an intersection product can be easily developped. This will be done in §2. The next task would be to prove the analogues of the main theorems of classical surface theory as Riemann-Roch, Hodge index theorem and Noether's formula.

For example the Riemann-Roch theorem classically for the case of an algebraic surface says:

$$\chi(\mathcal{O}_X(D)) - \chi(\mathcal{O}_X)$$
$$= \frac{1}{2} <D, D-\omega_X> ,$$

ω_X the canonical class.

Now the intersection number on the right in our case involves the infinite primes $v \in S_\infty$ of B , so should the left side. We consider the cohomology groups $H^i(X, \mathcal{O}_X(D))$, $i=0,1$, which are finitely generated R-modules.

Now suppose for a moment, we are in the classical situation of a fibration $\pi : X \to B$ of a surface over a curve B and would extend everything to the complete curve $\bar{B}$, $S_\infty = \bar{B} \setminus B$ the primes at infinity. If $\eta \in B$ is the generic point of B, $X_\eta = \pi^{-1}(\eta)$ the generic fibre, this would induce on the K-vector spaces $H^i(X_\eta, \mathcal{O}_X(D)|_{X_\eta})$ v-adic structures using the (canonical) isomorphisms

$$H^i(X_\eta, \mathcal{O}_X(D)|_{X_\eta}) \cong H^i(X_v, \mathcal{O}_X(D)|_{X_v}) \otimes_{R_v} K$$

for $v \in S_\infty$, $X_v = \pi^{-1}(\mathrm{Spec}(R_v))$.

Therefore, making use of the general philosophy, we could expect hermitian structures on the $(H^i(X, \mathcal{O}_X(D)) \otimes_R K)$ in our situation at all the infinite primes $v \in S_\infty$. Actually this seems to be hoping to much. What can be done is only to construct volume forms for $v \in S_\infty$, not even on the $H^i(X_v, O_X(D)|_{X_v})$, but on $H^0(X_v, \mathcal{O}_X(D)|_{X_v}) - H^1(X_v, \mathcal{O}_X(D)|_{X_v})$, that is, more precisely, a hermitian metric on

$$\lambda(H^0(X_v, \mathcal{O}_X(D)|_{X_v}) \otimes \lambda(H^1(X_v, \mathcal{O}_X(D)|_{X_v})^{-1}$$

where λ always denotes the highest non trivial exterior product.

Using this Faltings is able to prove in [F] all the analogues of the mentioned results of classical surface theory

In this paper we will do the following: We will introduce the intersection theory as well as the volume forms on $\chi(\mathcal{O}_X(D))$ in complete detail. Hopefully this is of help to algebraists which had so far not much experience with hermitian "analytic geometry". Afterwards we prove the Riemann-Roch as well as the Hodge index theorem, which both are fairly easy to obtain. We omit the proof of M.Noether's theorem, which is substantially deeper. We also omit the interesting considerations concerning the Arakelov Zeta functions as well as the explicit computations in the case of an elliptic curve. For all of this we refer the reader to Falting's paper [F].
One final comment: It would be nice to have volume forms also in the case of vector bundles E on X, that is volume forms on

$$\lambda\, \mathbb{R}\,\Gamma(X,E) = \lambda(H^{0}(X,E) \otimes \lambda(H^{1}(X,E))^{-1}$$

Apparently D. Quillen has results in this direction working in a more analytic context with Selberg's Zeta function, analytic torsion etc. We discuss this point a little bit at the end of §3 .

I would like to thank G. Faltings for explaining to me a number of points concerning his work.

§ 1 HERMITIAN LINE BUNDLES

(general reference for things not made explicit is the book of Griffiths and Harris, [G-H]).

We consider a Riemann surface X with genus $g > 0$. On the space of holomorphic differentials $\Gamma(X, \Omega_X^1)$ we have the hermitian form

$$<\omega_1, \omega_2> := \frac{i}{2} \int_X \omega_1 \wedge \bar{\omega}_2$$

Denote $\omega_1, \ldots, \omega_g$ an orthonormal basis of X.

We have the volume form

$$d\mu = \frac{i}{2g} \sum_{j=1}^{g} \omega_j \wedge \bar{\omega}_j ,$$

such that in particular $\int_X d\mu = 1$. $d\mu$ is independent of the orthonormal basis chosen.

Suppose, $\mathcal{L}$ is a hermitian line bundle on X, with metric $\| \ \|$. Canonically attached to $\mathcal{L}$ is its curvature form

$$\begin{aligned} \operatorname{curv}_{\mathcal{L}, \| \|} &:= \partial\bar{\partial} \log \|s\|^2 \\ &= \frac{\partial^2}{\partial z \partial \bar{z}} \log \|s\|^2 \quad dz \wedge d\bar{z} \end{aligned}$$

in local coordinates, where s is a meromorphic section of $\mathcal{L}$.

Apparently, the 1-1-form $\operatorname{curv}_{\mathcal{L}}$ is independent of the chosen section and therefore in particular well defined, because to any point $P \in X$ one can choose a section s, generating $\mathcal{L}$ in a neighborhood of P and compute $\mathrm{curv}_{\mathcal{L}}$ using this section there.

Remark: The definition of course makes sense for any complex manifold with hermitian line bundle on it.

The following is well known or can be easily derived using Stokes theorem.

Theorem 1: One has

$$\int_X \mathrm{curv}_{\mathcal{L}} = (2\pi i)\deg(\mathcal{L})$$

Therefore not any 1-1-form ω can occur as curvature form of a specified line bundle $\mathcal{L}$. On the other hand we will see below, that this relation above is the only obstruction to solving the equation

$$\mathrm{curv}_{\mathcal{L},\|\ \|} = \omega$$

We have to make use of

Proposition 1: Suppose, X is a Kähler manifold, η a 1-1-form, such that

a) $d\eta = 0$

b) η is perpendicular to the harmonic 1-1-forms with respect to the pairing given by the Kähler structure.

Then $\eta = \partial\bar{\partial}(v)$ can be solved with a C^∞-function v.
Furthermore v is uniquely determined up to a constant.

Proof: Using Hodge theory (with respect to d), we can write

$$\eta = h + d\eta_1 + d^*\eta_2 ,$$

an orthogonal decomposition, with h harmonic, d^* adjoint to d.

Because $d\eta=0$, we obtain $dd^*(\eta_2)=0$, therefore $d^*(\eta_2)=0$.
Using

$$0 = (\eta,h) = (h,h) \quad \text{by b)}$$

we have $h=0$.

Write $\eta_1=\eta_{1,0}+\eta_{0,1}$, $\eta_{1,0}$ a 1-0-form, $\eta_{0,1}$ a 0-1-form.
But because $\partial(\eta_{1,0})$ would be a 2-0-form, which could not cancel in

$$\eta = d\eta_1 = \partial\eta_1 + \bar{\partial}\eta_1 ,$$

we obtain $\partial(\eta_{1,0})=0$, as well as
$\bar{\partial}(\eta_{0,1})=0$

Using Hodge theory again (this time with respect to ∂,∂^* resp. $\bar{\partial},\bar{\partial}^*$), we can write

$$\left.\begin{aligned} \eta_{1,0} &= h_{1,0} + \partial\eta_{0,0} \\ \eta_{0,1} &= h_{0,1} + \bar{\partial}\eta_{0,0}^{(1)} \end{aligned}\right\}$$

where $h_{1,0}$ is harmonic with respect to ∂ and ∂^* , $h_{0,1}$ with respect to $\bar{\partial},\bar{\partial}^*$. Putting $v:=(-\eta_{0,0}+\tilde{\eta}_{0,0}^{(1)})$, we obtain $\partial\bar{\partial}(v)=\eta$ as a solution.
The uniqueness up to a constant follows (with a little care) from the maximum principle for harmonic functions.

Proposition 1 has several applications..

I) <u>Theorem 2:</u> Given a 1-1-form ω on the Riemann surface X which satisfies

$$\int_X \omega = (2\pi i)\ \deg(\mathcal{L})$$

Then there exist a hermitian metric $\|\ \|$ on $\mathcal{L}$, such that $\mathrm{curv}_{\mathcal{L},\|\ \|} = \omega$. $\|\ \|$ is determined up to a positive constant factor.

Proof: Choose an arbitrary hermitian metric $\|\ \|_1$ on $\mathcal{L}$; Suppose

$$\mathrm{curv}_{\mathcal{L},\|\ \|_1} = \omega_1$$

By theorem 1 we have $\int_X (\omega-\omega_1)=0$. $(\omega-\omega_1)$ certainly is closed. The space of harmonic 1-1-forms is 1-dimensional, generated by $d\mu$, furthermore

$$(\omega-\omega_1, d\mu) = \int_X (\omega-\omega_1) = 0$$

By Propos-ition 1 we can solve $\partial\bar{\partial}(v)=(\omega-\omega_1)$. Putting $\exp(v/_2)=:u$, we can define

$$\| \ \| \ := u \ \| \ \|_1$$

and obtain a hermitian metric with curvature form ω . The uniqueness up to aconstant factor follows as above. q.e.d.

As an immediate application of this we obtain a uniquely determined hermitian metric on any line bundle $\mathcal{L}$ on the Riemann surface X as follows:

i) Suppose first, $Q \in X$, $\mathcal{L}=\mathcal{O}_X(Q)$. Then there is a uniquely determined metric $\| \ \|$ on $\mathcal{L}$, such that for

$$G(P,Q) := \|1\|_{\mathcal{O}_X(Q)}(P) ,$$

the length of the constant section $1 \in \Gamma(X,\mathcal{O}_X(Q))$ at P , we have

a) $\partial_P \bar{\partial}_P \log G^2(P,Q)$

$$= -\frac{\pi}{g}\left(\sum_{j=1}^{g} \omega_j \wedge \bar{\omega}_j\right)$$

b) $\int_X \log G(P,Q)\, d\mu(P)=0$

ii) Writing an arbitrary line bundle as a tensor product of $\mathcal{O}_X(Q)$'s , we obtain a uniquely determined hermitian metric on any $\mathcal{L}$ on X .

We call a hermitian metric $\|\ \|$ on $\mathcal{L}$, with $\mathrm{curv}_{\mathcal{L},\|\ \|} = c\, d\mu$ $c=$ constant, admissible. Making use of the extra condition b) we have specified a unique admissible metric.

Remarks: 1) We pose $g(P,Q):=\log G(P,Q)$. g is a C^{∞}-function for all $P\neq Q$. The behavior at $P=Q$ is as follows: locally around Q we can write

$$1 = z\cdot s \quad ,$$

s a generating section of $\mathcal{O}_X(Q)$ in Q, z a local coordinate around Q. Therefore $\|1\| = |z|\cdot\|s\|$, where $\|s(Q)\|\neq 0$, hence

$$g(P,Q) = \log|z(P)| + \log\|s(P)\|$$

and $g(P,Q)$ has a logarithmic singularity at $P=Q$.

Remark: $-g(P,Q)$ gives an inverse (Green function) for the positive elliptic differential oparator Δ defined by

$$\partial\bar{\partial}(f) = -\frac{\Delta(f)}{2g}\cdot\pi\cdot\sum_{j=1}^{g}\omega_j\wedge\bar{\omega}_j$$

For details, see [F]

2) One should remark, that $\int_X g(P,Q)\, d\mu(P)$ exists, the singularity in Q causes no difficulties ($\int_0^{\epsilon} r\log r\, dr$ exists!)

More generally we have

<u>Theorem 3</u>: Suppose, X is a Kähler manifold, ω a 1-1-form on X, $\mathcal{L}$ a line bundle on X. Then the equation

$$\mathrm{curv}_{\mathcal{L},\|\ \|} = \omega$$

can be solved, iff

1) $d\omega = 0$
2) $[\omega]$, the cohomology class represented by ω, satisfies
   $[\omega] = 2\pi i \ c_1(\mathcal{L})$

The proof is similar to the proof of theorem 2 and can be found in |G-H| ,p. 139-144. (But caution: Griffith uses $\mathrm{curv}_{\mathcal{L}} = \bar{\partial}\partial \log \|\ldots.\|$, hence a (-) sign!) There are other possibilities to express property 2) . For example, suppose $\mathcal{L} = \mathcal{O}_X(D)$, $D = \Sigma\, n_i Y_i$, where the Y_i are (n-1)-dimensional subvarieties. Then

$$\int_X \omega \wedge h = 2\pi i \left(\sum_i n_i \int_{Y_i} h \right)$$

should hold for all harmonic (n-1)-(n-1)-forms h.

We can apply this in the following case: Consider (for a Riemann surface X) the Kähler manifold $X \times X$ and the line bundle $\mathcal{L} = \mathcal{O}_{X\times X}(\Delta(X))$, $\Delta(X)$ the diagonal.

Take

$$\omega = 2\pi i(p_1^* d\mu + p_2^* d\mu)$$

$$- \pi \sum_{j=1}^{g} (p_1^*(\omega_j) \wedge p_2^*(\bar{\omega}_j) + p_1^*(\bar{\omega}_j) \wedge p_2^*(\omega_j))$$

p_1, p_2 of course the projections.

Checking against a generating system of harmonic 1-1-forms, as for example $p_1^* d\mu$, $p_2^* d\mu$, $p_1^*(\omega_i) \wedge p_2^*(\bar{\omega}_j)$, $p_1^*(\bar{\omega}_i) \wedge p_2^*(\omega_j)$ condition b) (or better the equivalent version) above, we easily obtain.

<u>Theorem 3:</u> There is a unique hermitian metric $\|\ \|$ on $\mathcal{L}$ such that

a) $\mathrm{curv}_{\mathcal{L}, \|\ \|} = \omega$

b) $\int_X \log \| 1 \| (P,Q)\, d\mu(P) = 0$

for $Q = Q_o \in X$ a specified point.

II) We determine the relation of the function $\| 1 \| (P,Q)$ on $(X \times X)$ to our previously considered function $G(P,Q)$.

As ω is symmetric, we have

$$\| 1 \| (P,Q) = c \cdot \| 1 \| (Q,P) \ ,$$

$$0 < c \in \mathbb{R} \ .$$

Therefore $c=1$, applying this twice.

We will show in a moment, that $\phi(Q) := \int_X \log \| 1 \| (P,Q)\, d\mu(P)$ is a constant function. Therefore $\phi(Q) \equiv 0$ by b).

But then, by restricton, we obtain

<u>Supplement to Theorem 3:</u> One has $\| 1 \| (P,Q)=G(P,Q)$, in particular $G(P,Q)$ and $g(P,Q)$ are symmetric functions

It remains to show

<u>Lemma1:</u> The function

$$\phi(Q) = \int_X \log \| 1 \| (P,Q)\, d\mu(P)$$

is constant.

Proof: We compute

$$\partial_Q \bar{\partial}_Q \int_X \log \| 1 \| (P,Q)\, d\mu(P) \Big|_{Q_1} ,$$

Q_1 an arbitrary point of X .

Suppose $U_\epsilon(Q_1)$ is a small ϵ-neighborhood in X around Q_1 , $U_{\epsilon/2}(Q_1) \subset U_\epsilon(Q_1)$ and α_1, α_2 real valued positive C^∞-functions on X , such that

$$\begin{aligned} &\text{i)} \quad \operatorname{supp}(\alpha_1) \subset U_\epsilon(Q_1) \\ &\text{ii)} \quad \alpha_1 = 1 \text{ on } U_{\epsilon/2}(Q_1) , \\ &\qquad \alpha_2 = 1 \text{ on } X \setminus U_\epsilon(Q_1) \\ &\text{iii)} \quad \alpha_1 + \alpha_2 = 1 \text{ on } X . \end{aligned}$$

Therefore

$$\begin{aligned} &\partial_Q \bar{\partial}_Q \phi(Q) \Big|_{Q_1} \\ &= \lim_{\epsilon \to 0} \partial_Q \bar{\partial}_Q \int_{X \setminus U_{\epsilon/2}(Q_1)} \log \| 1 \| (P,Q)\, d\mu(P) \Big|_{Q_1} \\ &+ \lim_{\epsilon \to 0} \partial_Q \bar{\partial}_Q \int_X \alpha_1(P) \log \| 1 \| (P,Q)\, d\mu(P) \Big|_{Q_1} \end{aligned}$$

We obtain for the first term

$$\lim_{\epsilon\to 0} \int_{X\setminus U_{\epsilon/2}(Q_1)} \partial_Q \bar{\partial}_Q \log\| 1\| (P,Q)|_{Q_1} \, d\mu(P)$$

$$= \int_X \left(-\frac{\pi}{2g} \sum_{j=1}^{g} (\omega_j\wedge\bar{\omega}_j)(Q_1) d\mu(P)\right.$$

$$= (\pi i)\, d\mu(Q_1)$$

For the second term we can introduce local coordinates (t,z) for (P,Q) around Q_1 and obtain

$$\partial_z \bar{\partial}_z \int_{|t|\leq\epsilon} \log|z-t| \ \psi(t)d\mu(t)|_{z=0} \ ,$$

where

$d\mu(t)$ = the standard measure on $\mathbb{C}$,

$\psi(t)\ d\mu(t) = d\mu(P)$ on $U_{\epsilon/2}(Q_1)$,

corresponding some open neighborhood of $t=0$, finally $\psi(t)$ with compact support in $|t|<\epsilon$.

We can write

$$\partial_z\bar{\partial}_z \int_{|t|\leq\epsilon} \log|z-t|\psi(t)d\mu(t)|_{z=0}$$

$$=\partial_z\bar{\partial}_z \int_{\mathbb{C}} \log|z-t|\psi(t)d\mu(t)|_{z=0}$$

$$=\partial_z\bar{\partial}_z \int_{\mathbb{C}} \log|u| \ \psi(z+u)d\mu(u)|_{z=0} \qquad (t-z=:u)$$

$$= \int_{\mathbb{C}} \log|u|\,\partial_z\bar{\partial}_z \ \psi(z+u)|_{z=0}\, d\mu(u)$$

$$= \left(\int_{\mathbb{C}} \log|u| \left(\frac{1}{4}\right) \Delta_u (\psi(u)) \, d\mu(u) \right) dz \wedge d\bar{z}$$

$$= \left(\lim_{\delta \to 0} \int_{\delta \le |u| \le R} (\log|u|) \, \frac{1}{4} \Delta_u (\psi(u)) \, d\mu(u) \right) dz \wedge d\bar{z}$$

$$= \Big(\lim_{\delta \to 0} \int_{|u|=\delta} \frac{1}{4} \, \frac{\partial}{\partial n} (\log|u|) \, \psi(u) \, ds$$

$$- \int_{|u|=\delta} \frac{1}{4} \log|u| \, \frac{\partial \psi(u)}{\partial n} \, ds \Big) \quad dz \wedge d\bar{z}$$

using Green's theorem.

$$= \left(\lim_{\delta \to 0} \int_{|u|=\delta} \frac{1}{4|u|} \, \psi(u) ds \right) dz \wedge dz$$

$$= \frac{\pi}{2} \, \psi(0) \quad dz \wedge d\bar{z}$$

$$= - \pi i \quad \psi(0) \; d\mu(z)$$

$$= - \pi i \quad d\mu(Q_1) \; .$$

Therefore

$$\partial_Q \, \bar{\partial}_Q \, \phi(Q) = 0$$

and $\phi(Q)$ has to be constant. q.e.d.

§ 2 ARAKELOV-DIVISORS AND INTERSECTION THEORY

Suppose $B=\mathrm{Spec}(R)$, where R is the ring of algebraic integers in the number field K, $\pi : X \to B$ a semistable curve over B, $\eta \in B$ the generic point, S_f the set of closed points of B (finite places of K), S_∞ the set of infinite places, $S=S_f \cup S_\infty$,

$$X_v := X_\eta \otimes_K \hat{K}_v \quad \text{for } v \in S_\infty$$

the associated Riemann surfaces for the infinite primes.

<u>Definition 1:</u> The group of Arakelov.divisors is

$$\widetilde{\mathrm{Div}}(X) = \mathrm{Div}(X) \oplus \bigoplus_{v \in S_\infty} \mathbb{R}(X_v)$$

So, any Arakelov-divisor has a unique decomposition

$$D = D_f + D_\infty ,$$

where $$D_\infty = \sum_{v \in S_\infty} r_v (X_v)$$

Now using the results of §1, we can associate with any Arakelov-divisor D a set of hermitian line bundles for the $v \in S_\infty$. For a fixed v, the line bundle itself will be the one induced from $\mathcal{O}_X(D_f)$ on X to X_v.
This line bundle has a canonical hermitian metric by the results of §1. To take into account the infinite part D_∞ of D, the hermitian metric has to be rescaled by the factor

$\exp(-r_v)$.

<u>Definition 2:</u> By a <u>hermitian line bundle</u> on the arithmetic surface X , associated to the Arakelov-divisor D we understand the line bundle $\mathcal{O}_X(D_f)$, enriched with the hermitian metrics at the $v \in S_\infty$, explained above.

We remind the reader that a hermitian line bundle on $B=\mathrm{Spec}\,(R)$ has a degree, given as follows. The line bundle is given by a projective module P of rang 1 over R , suppose $p \in P$, $p \neq 0$: Then we have

$$\deg(P) = \log \#(P_{/Rp}) - \sum_{v \in S_\infty} \epsilon_v \log \|p\|_v$$

where

$$\epsilon_v = \begin{cases} 1, & \text{if } \hat{K}_v = \mathbb{R} \\ 2, & \text{if } \hat{K}_v = \mathbb{C} \end{cases}$$

<u>Definition of the intersection product:</u>

This will be uniquely determined by the following properties of the intersection pairing

$$\left.\begin{array}{ccc} \widetilde{\mathrm{Div}}(X) \times \widetilde{\mathrm{Div}}(X) & \to & \mathbb{R} \\ (D_1,D_2) & \to & \langle D_1,D_2\rangle \end{array}\right\}$$

1.) $\langle D_1,D_2\rangle$ is biadditive

2.) Suppose, D_1 is an irreducible vertical ($\subseteq \pi^{-1}(v), v \in S$) divisor .

Then

$$\langle D_1, D_2\rangle = \deg(\mathcal{O}_X(D_2)|_{D_1}) \cdot \begin{cases} \log(q_v), & v \in S_f \\ 1, & S_\infty \end{cases}$$

, where $q_v = \# k(v)$ the order of the residue field.

3.) Suppose $L \supset K$ is a finite field extension, X_L a semistable regular model of $(X_\eta \otimes_K L)$, $\phi : X_L \to X_K$ the projection. Then one has

$$\langle D_1, D_2\rangle_{X_K} = \frac{1}{(L:K)} \langle D_1, D_2\rangle_{X_L}$$

4.) Suppose $D_1 = (P)$, $P \in X_\eta(K)$ a rational point. P defines a section $s : B \to X$. $\mathcal{O}_X(D_2)$ is a hermitian line bundle on X, therefore $s^*(\mathcal{O}_X(D_2))$ is a hermitian line bundle on B and one has

$$\langle D_1, D_2\rangle = \deg s^*(\mathcal{O}_X(D_2))$$

<u>Remarks:</u> It is easy to check, that properties 1.) - 4.) uniquely determine the intersection pairing.
We now have to establish the usual properties of an intersection pairing, that is:

<u>Theorem 1:</u>
1.) If D_1, D_2 have no common components, $\langle D_1, D_2\rangle$ can be determined by computing <u>local intersection numbers</u>.

2.) $\langle D_1, D_2\rangle = \langle D_2, D_1\rangle$

3.) Suppose $f \in K(X)$ is a function, (f) the associated divisor on X,

$$(f)^{\sim} := (f) + \sum_{v \in S_\infty} r_v(X_v)$$

with

$$r_v := \; - \int_{X_v} \log \| f \| \, d\mu_v \qquad \text{for } v \in S_\infty$$

the <u>Arakelov-divisor associated to f</u> .

Then one has

$$\langle (f)^{\sim}, D \rangle = 0$$

for all $D \in \mathrm{Div}(X)$.

<u>Proof:</u> It is enough to show 1.) and 3.) , because using 3.) we can always assume D_1, D_2 without common components. Because we will see that the local intersection numbers are symmetric, 2.) follows.

We show 1.) , but only for the typical case $(P) = D_1$, $P \in X_\eta(K)$ a rational point. We can assume, that D_2 is an effective divisor on X . We consider the section $p = 1 \in \Gamma(X, \mathcal{O}_X(D_2))$

We have

$$\begin{aligned} \langle D_1, D_2 \rangle &= \deg(\mathcal{O}_X(D_2) \,|\, _{(P)} \\ &= \left. \begin{aligned} &\log \#(\mathcal{O}_X(D_2) \,|\, _{(P)} \,/\, R\, 1 \\ &- \sum_{v \in S_\infty} \epsilon_v \log \| 1 \|_v \end{aligned} \right\} \end{aligned}$$

Suppose $x \in D_1 \cap D_2$, $\pi(x) = v \in B$, $t=0$ and $z=0$ local equations for D_1, D_2 in x. We have

$$\mathcal{O}_X(D_2)_{(x)} = z^{-1} \mathcal{O}_{X,x} \supset 1 \cdot \mathcal{O}_{X,x} \,,$$

furthermore

$$0 \to (t) \to \mathcal{O}_{X,x} \to \mathcal{O}_{D_1,x} \to 0$$

Using the isomorphism

$$(z^{-1}\, \mathcal{O}_{D_1,x} / 1 \cdot \mathcal{O}_{D_1,x}) \xrightarrow{\sim} \mathcal{O}_{X,x} / (t,z)$$

we obtain for the local contributions

$$\begin{aligned} \langle D_1, D_2 \rangle_{(x)} &= \log \#(\mathcal{O}_{X,x}/(t,z)) \\ &= (\log(q_v)) \cdot (D_1, D_2)_x \quad , \end{aligned}$$

$(D_1, D_2)_{(x)}$ the usual intersection multiplicity of D_1, D_2 at x .

As x and $\pi(x)=v$ uniquely determine each other, we will write also $\langle D_1, D_2\rangle_v$ for these contributions. There remain the contributions at $v \in S_\infty$.

Write $D_2 = D_2^h + D_2^v$, a sum of horizontal and vertical divisors. On X_v, $v \in S_\infty$, we have

$$D_2^{(h)} = \Sigma\, n_Q (Q)$$

Therefore we obtain for these $v \in S_\infty$

$$\begin{aligned} \langle D_1, D_2 \rangle_v &= -\log \|1\| (P) \\ &= -\Sigma\, n_Q \log G^{(v)}(P,Q) \end{aligned}$$

as a local expression. We see again, that $\langle D_1, D_2\rangle_v$

is symmetric, because the functions $G^{(v)}(P,Q)$ on the X_v are symmetric.

Altogether we obtain

$$<D_1,D_2> = \sum_{v \in S} <D_1,D_2>_v ,$$

a decomposition of the intersection number in local intersection numbers.

We next show 3.) of theorem 1: It suffices again to do the case $D_1=(P)$, $P \in X_\eta(K)$ a rational point. That is, we have to show

$$<P,(f)^\sim> = 0$$

Consider the hermitian line bundle $O_X((f)^\sim)$ on X. We take $f^{-1}=p$ as a section and obtain

$$
\begin{aligned}
<(P),(f)^\sim> &= \deg(s^*\mathcal{O}_X((f)^\sim)) \\
&= \log \# s^*(\mathcal{O}_X((f)^\sim) / f^{-1}R - \sum_{v \in S_\infty} \epsilon_v \log \| f^{-1} \|_v \\
&= - \sum_{v \in S_\infty} \epsilon_v \log \| f^{-1} \|_v
\end{aligned}
$$

As $(f)=\Sigma n_Q(Q)$ on X_v, we can write

$$|f| = \prod_Q G(P,Q)^{n_Q} u(P) ,$$

$u(P)$ a C^∞-function on X, $u(P) \neq 0$ on X.

Therefore we obtain

$$
\begin{aligned}
0 &= \partial\bar\partial \log |f| \\
&= \sum_Q n_Q \; \partial\bar\partial \log G(P,Q) + \partial\bar\partial \log u(P) \\
&= \partial\bar\partial \log u(P) \;, \text{ because } \sum n_Q = 0
\end{aligned}
$$

It follows, that $u(P)=c_v$ is constant.

Because

$$\int_X \log G(P,Q)\, d\mu(P) = 0 \;, \quad \text{we obtain}$$

$$
\begin{aligned}
\int_{X_v} \log |f|\, d\mu &= \int_{X_v} \log u(P)\, d\mu(P) \\
&= (\log c_v) \int_X d\mu(P) = \log (c_v)
\end{aligned}
$$

Finally

$$
\begin{aligned}
\log \| f^{-1} \|_v &= \log \| f^{-1} \cdot 1 \|_v \\
&= \log |f^{-1}| + \log \| 1 \|_v \\
&= (\sum_Q -n_Q \log G(P,Q) - \log u(P)) \\
&\quad + \sum_Q n_Q \log G(P,Q) + (-r_v) \\
&= - \int_{X_v} \log |f| d\mu + \int_{X_v} \log |f|\, d\mu \\
&= 0 \qquad\qquad \text{q.e.d.}
\end{aligned}
$$

§ 3 VOLUME FORMS ON $\mathbb{R}\Gamma(X,\mathcal{L})$

As explained in the introduction to be able to formulate theorems like the Riemann-Roch theorem for arithmetic surfaces it is necessary to have at least a volume form on the virtual R-module

$$H^0(X;\mathcal{L}) - H^1(X;\mathcal{L}) \ ,$$

that is, more precisely, a hermitian structure on the $\hat{K}_v$-vector spaces

$$\lambda(H^0(X,\mathcal{L}) \otimes_R \hat{K}_v) \otimes \lambda(H^1 X,\mathcal{L}) \otimes_R \hat{K}_v)^{-1}$$

where λ denotes the highest non trivial exterior product of the $\hat{K}_v$-vektorspace $H^0(X,\mathcal{L}) \otimes_R \hat{K}_v$ for example $(v \in S_\infty)$
We will handle this problem in the context of Riemann surfaces, so let X be again a Riemann surface of genus $g \geq 1$ in this paragraph and we use the same notations as §1 .

Definition 1: We put formally

$$\lambda(H^0(X,\mathcal{L})) \otimes \lambda(H^1(X,\mathcal{L}))^{-1} =: \lambda\, \mathbb{R}\Gamma(X,\mathcal{L})$$

for a line bundle $\mathcal{L}$ on X .

We consider only such hermitian metrics on a line bundle $\mathcal{L}$ such that the curvature form $\mathrm{curv}_{\mathcal{L},\|\ \|}$ is a multiple of $d\mu = \left(\frac{i}{2g}\right) \Sigma\, (\omega_j \wedge \bar{\omega}_j)$.
Denote $\underline{\mathcal{C}}$ the category of all such hermitian line bundles on X with isometries.

<u>Theorem 1:</u> There is a functor $\mathcal{L} \mapsto \lambda\, \mathbb{R}\Gamma(X,\mathcal{L})$ on the category $\underline{\mathcal{C}}$ to the category of hermitian (1-dimensional) complex vector spaces, such that the following properties hold:

i) Any isometric isomorphism $\mathcal{L}_1 \to \mathcal{L}_2$ in $\underline{\mathcal{C}}$ induces an isometric isomorphism $\lambda\mathbb{R}\Gamma(X,\mathcal{L}_1) \to \lambda\,\mathbb{R}\Gamma(X,\mathcal{L}_2)$ that is, saying again, $\lambda\,\mathbb{R}\Gamma(X,?)$ is a functor.

ii) If one changes the metric on a line bundle $\mathcal{L}$ by a factor $\alpha > 0$, the metric on $\lambda\mathbb{R}(\Gamma(X,\mathcal{L}))$ changes by the factor $\alpha^{\chi(\mathcal{L})} = \alpha^{h^0(\mathcal{L}) - h^1(\mathcal{L})}$, $h^i(\mathcal{L}) : \dim_{\mathbb{C}} H^i(X,\mathcal{L})$ (i=1,2).

iii) Suppose, D is a divisor on X, $P \in X$, $D_1 = D - P$ and $\mathcal{O}_X(D_1)$, $\mathcal{O}_X(D)$ are equipped with the hermitian structure introduced in §1. The one-dimensional fibre of $\mathcal{O}_X(D)$ at P, $\mathcal{O}_X(D)[P]$ inherits the hermitian vector space structure of $\mathcal{O}_X(D)$. Then the <u>canonical map</u>

$$\begin{aligned} &\lambda\mathbb{R}\Gamma(X,\mathcal{O}_X(D))) \\ &\cong \lambda(\mathbb{R}\Gamma(X,\mathcal{O}_X(D_1)))\otimes_{\mathbb{C}}\mathcal{O}_X(D)[P] \end{aligned}$$

is an isometric isomorphism. The functor $\lambda\mathbb{R}\Gamma(X,\ )$ is uniquely determined by i) - iii) up to a factor > 0.

<u>Remark:</u> Suppose $0 \to V_1 \to V_2 \to .. \to V_n \to 0$ is an exact sequence of vector spaces. Then there is a canonical homomorphism

$$\left.\begin{aligned} &\lambda(V_1) \otimes \lambda(V_3) \otimes \\ \xrightarrow{\sim}\ &\lambda(V_2) \otimes \lambda(V_4) \otimes \end{aligned}\right\}$$

This ismorphism is the one meant above.

Proof: A) We first construct an assignment

$$D \to \lambda \mathbb{R}\Gamma(X,\mathcal{O}_X(D))$$

associating to any divisor D with a specified admissible hermitian metric on $\mathcal{O}_X(D)$ a volume form on $\mathbb{R}\Gamma(X,\mathcal{O}_X(D))$, such that this map by construction fullfills ii) and iii): This is in fact easy . Fix any volume form on $\mathbb{R}\Gamma(X,\mathcal{O}_X))$, that is for $D=0$. Next build up $\mathcal{O}_X(D)$ by adding and subtracting points. Property iii) (and ii)) say how to define $\lambda\mathbb{R}\Gamma(X,\mathcal{O}_X(D))$ in general. The fact, that the functions $G(P,Q)$ are symmetric, guarantees, that it plays no role, how D is build up from nothing. There remains to show, that the map $D \mapsto \lambda\mathbb{R}(X,\mathcal{O}_X(D))$ in fact induces a functor on $\underline{\underline{\mathfrak{C}}}$, that is to prove i)

B) Proof of i): Suppose we have two divisors D,D' and an isometry $\mathcal{O}_X(D) \to \mathcal{O}_X(D')$. To show: The induced map

$$\lambda\mathbb{R}\Gamma(X,\mathcal{O}_X(D)) \to \lambda\mathbb{R}\Gamma(X,\mathcal{O}_X(D'))$$

is an isometry itself.
It suffices to show this for divisors with a specified degree, making use of iii) again.

Suppose therefore,

$$\deg(D) = \deg(D') = (g-1) .$$

We can write

$$\left.\begin{aligned} D &= E-(P_1+\ldots+P_r) \\ D' &= E-(P'_1+\ldots+P'_r) \end{aligned}\right\}$$

with an effective divisor E , if r is large enough.

Consider the map

$$\varphi: X^r \to \mathrm{Pic}_{g-1}(X)$$
$$(P_1,\ldots,P_r) \mapsto \mathcal{O}_X(E-\sum_{i=1}^{r} P_i)$$

The study of this map φ will enable us to prove i) . Using the standard properties of base change, it is easy to see, that we have a (hermitian) line bundle $\mathcal{H}$ on X^r , the fibre at $(P_1,\ldots,P_r)$ beeing $\lambda \mathbb{R}\Gamma(X,\mathcal{O}_X(E-\sum_{i=1}^{r} P_i))$ On the other hand, on $\mathrm{Pic}_{g-1}(X)$ we have the theta-divisor $\theta = \mathrm{image}(X^{g-1} \to \mathrm{Pic}_{g-1}(X))$. The associated line bundle $\mathcal{O}(-\theta)$ on $\mathrm{Pic}_{g-1}(X)$ will obtain a hermitian structure and we will show, that the pull-back of $\mathcal{O}(-\theta)$ as a hermitian line bundle will be $\mathcal{H}$ up to a constant factor. Therefore it follows, that the volume element on $\mathbb{R}\Gamma(X,\mathcal{O}_X(E - \sum_{i=1}^{r} P_i))$ depends in fact only on the isomorphism class of $\mathcal{O}_X(E - \sum_{i=1}^{r} P_i)$. Using this, i) of theorem 1 follows.

We therefore have to fullfill the following program:

a) Construct a hermitian metric on $\mathcal{O}(-\theta)$.

b) To show: $\varphi^*(\mathcal{O}(-\theta))$ and $\mathcal{H}$ are isometric up to a factor. To see this, it suffices to show:

$$\varphi^*(\mathrm{curv}_{\mathcal{O}(-\theta)}) = \mathrm{curv}_{\mathcal{H}}$$

Ad a): Using an embedding $X \longrightarrow \mathrm{Pic}_{g-1}(X)$, we can identify the differential forms $\omega_1,\ldots,\omega_g$ with forms on $\mathrm{Pic}_{g-1}(X)$ which we also denote by $\omega_1,\ldots,\omega_g$ and which are translation-invariant.

Proposition 1: There is a hermitian metric on $\mathcal{O}(-\theta)$, such that the curvature form is

$$\pi \sum_{j=1}^{g} (\omega_j \wedge \bar{\omega}_j) =: \eta$$

Indication of proof: Of course we want to use theorem 3 of §1 . That is, we have to show for h an arbitrary harmonic $(g-1,g-1)$ form, that

$$\int_{\mathrm{Pic}_{g-1}(X)} (\eta \wedge h) = -(2\pi i) \int_{\theta} h$$

That is, the 1-1-form $(-2\pi i\eta)$ represents the cohomology class associated to θ . It is enough to check this for a generating system of harmonic $(g-1)-(g-1)$ forms, for example for the forms $\omega_I \wedge \bar{\omega}_J$, where $\omega_I = \bigwedge_{i\in I} \omega_i$, $\bar{\omega}_J = (\bigwedge_{j\in J} \bar{\omega}_j)$, $\#I = \#J = (g-1)$.

Finally one should evaluate the integrals involved as follows. Using the canonical map

$$X^{g-1} \overset{\psi}{\rightarrow} \theta \subset \mathrm{Pic}_{g-1}(X)$$

we have, because generically the map is

finite of degree (g-1) ! ,

$$\int_\theta h = \frac{1}{(g-1)!} \int_{X^{g-1}} \psi^*(\omega_I \wedge \bar{\omega}_J) ,$$

but the pull-backs $\psi^*(\omega_I)$ on X^{g-1} are easily determined. Similarly one proceeds with $\int_{\mathrm{Pic}_{g-1}(X)} (\eta \wedge h)$ and the map covering

$$\left.\begin{array}{l} X^g \overset{\psi}{\to} \mathrm{Pic}_{g-1}(X) \\ (Q_1,\ldots,Q_g) \longrightarrow (\sum_{j=1}^{g} Q_j - P_o) \end{array}\right\} \quad P_o \text{ a fixed point on } X ,$$

such that

$$\int_{\mathrm{Pic}_{g-1}(X)} (\eta \wedge h) = \frac{1}{g!} \int_{X^g} \psi^*(\eta \wedge h)$$

It is an easy exercise now to complete the proof of proposition 1 .

Ad b1): To show $\varphi^*(\mathcal{O}(-\theta)) \cong \mathcal{H}$ as line bundles (without hermitian structure for the moment):

One has to go back to the construction of the line bundles involved. On $(X \times X^r)$ respectively $(X \times \mathrm{Pic}_{g-1}(X))$ we have the obvious universal line bundles, say $\widetilde{\mathcal{H}}$ and $\mathcal{P}$. We have the diagram

$$\begin{array}{ccc} X \times X^r & \xrightarrow{(id,\varphi)} & X \times \mathrm{Pic}_{g-1}(X) \\ \downarrow \pi_2 & & \downarrow \pi_2 \\ X^r & \xrightarrow[\varphi]{} & \mathrm{Pic}_{g-1}(X) \end{array}$$

Obviously

$$(\mathrm{id},\varphi)^*(\mathcal{P}) \cong \tilde{\mathcal{H}}.\otimes \pi_2^*(\mathcal{L}_o)$$

with some line bundle $\mathcal{L}_o$ on X^r. Furthermore it is known, that

$$\mathcal{P} \cong \mathcal{O}(-\theta) .$$

One should remark for this, that $\mathcal{P}$ is trivial on $\varphi^{-1}(\mathrm{Pic}_{g-1}(X)\setminus\theta)$.

We obtain

$$\left.\begin{array}{l} \lambda\mathbb{R}(\pi_2)_*\ (\mathcal{P}) = \mathcal{O}(-\theta) \\ \lambda\mathbb{R}(\pi_2)_*\ (\tilde{\mathcal{H}}) = \mathcal{H} \end{array}\right\}$$

Using base change, we finally have the isomorphisms

$$\begin{array}{ccc} \varphi^*\lambda(\mathbb{R}(\pi_2)_*(\mathcal{P}) & \xrightarrow{\sim} & \lambda(\mathbb{R}(\pi_2)_*)(\varphi^*(\mathcal{P})) \\ \wr\downarrow & & \wr\downarrow \\ \varphi^*(\mathcal{O}(-\theta)) & & \mathcal{H} \end{array}$$

b1) follows. q.e.d.

We have

$$\left.\begin{array}{l} \mathcal{H} \cong \lambda\mathbb{R}\Gamma(X,\mathcal{O}(E)) \\ \otimes \ (\bigotimes_{i=1}^{r} p_i^*(\mathcal{O}_X(E))^{-1} \\ \otimes \bigotimes_{1\le k<l<r} p_{k,l}^*(\mathcal{O}_{X\times X}(\Delta(X))) \end{array}\right\}$$

Therefore

$$\left.\begin{aligned}\operatorname{curv}_{\mathcal{H}} &= -\sum_{i=1}^{r} p_i^{*}(\operatorname{curv}_{\mathcal{O}_X(E)}) \\ &\quad + \sum_{1\le k<l\le r} p_{k,l}^{*}(\operatorname{curv}_{\mathcal{O}_{X\times X}(\Delta(X))})\end{aligned}\right\}$$

But we have

$$\sum_{i=1}^{r} p_i^{*}(\operatorname{curv}_{\mathcal{O}_X(E)})$$

$$= \sum_{i=1}^{r} -\pi\cdot\left(\frac{\deg(E)}{g}\right) p_i^{*}\left(\sum_{j=1}^{g} (\omega_j\wedge\bar{\omega}_j)\right)$$

and

$$\sum_{1\le k<l\le r} p_{k,l}^{*}(\operatorname{curv}_{\mathcal{O}_{X\times X}}(\Delta(X)))$$

$$= \sum_{1\le k<l\le r} 2\pi i\,(p_k^{*}(d\mu) + p_l^{*}(d\mu)$$

$$- \sum_{1\le k<l\le r} \pi \sum_{j=1}^{g} (p_k^{*}(\omega_j)\wedge p_l^{*}(\bar{\omega}_j)+p_k^{*}(\bar{\omega}_j)\wedge p_l^{*}(\omega_j))$$

Taking this together, one obtains, using b1), by a short computation:

$$\operatorname{curv}_{\mathcal{H}} = \varphi^{*}(\operatorname{curv}_{\mathcal{O}(-\theta)})$$

(see [F] , if necessary)

Therefore the hermitian metrics on $\mathcal{H}$ and $\varphi^{*}(\mathcal{O}(-\theta))\simeq\mathcal{H}$ differ only by a constant. Theorem 1 follows. q.e.d.

<u>Remark:</u> As already mentioned in the introduction, it would be be interesting to do a similar thing for vector bundles on a Riemann surface X . If one wants to use the same method as followed here, one has the following problems:

(1) Specifying a curvature form for all bundles E of rang d in $(A^{1,1} \otimes \mathrm{End}(E))$ up to a multiple. Probably one should use $(\sum_{j=1}^{g} (\omega_j \wedge \bar{\omega}_j) \otimes \mathrm{Id})$, but perhaps this puts restrictions on the bundles E , indecomposable or stable perhaps.

(2) How to define the volume form

$$E \to \lambda \mathbb{R}\Gamma(X,E) \ ?$$

Even if one starts with a matrix divisor instead of a bundle it is not clear how to define $\lambda\mathbb{R}\Gamma(X,-)$, because a matrix divisor can be build up in many different ways.

§ 4 RIEMANN-ROCH

We consider again the arithmetic situation, that is, $\pi : X \to B$ our semistable curve over $B = \mathrm{Spec}(R)$, R the ring of algebraic integers in K .

Suppose, D is an Arakelov-divisor, $\mathcal{L} = \mathcal{O}_X(D)$ a hermitian line bundle on X ; Then $H^i(X;\mathcal{L}) = 0$ for $i \geq 2$, using the Leray spectral sequence and, using the results of §3, we have a volume form on the virtual $(R \otimes_{\mathbb{Z}} \mathbb{R})$-module

$$\left.\begin{aligned} &(H^0(X;\mathcal{L}) - H^1(X;\mathcal{L})) \otimes_{\mathbb{Z}} \mathbb{R} \\ &= \mathbb{R}\Gamma(X,\mathcal{L}) \otimes_{\mathbb{Z}} \mathbb{R} \end{aligned}\right\}$$

To be able to make computations, we develop the following formalism:

<u>Definition 1</u>: Suppose, M is a finitely generated R-module, vol a Haar measure on $(M \otimes_{\mathbb{Z}} \mathbb{R})$ (over $R \otimes_{\mathbb{Z}} \mathbb{R} \cong \prod_{v \in S_\infty} \hat{K}_v$)
Then one poses

$$\tilde{\chi}(M) := -\log\left(\frac{\mathrm{vol}(M \otimes_{\mathbb{Z}} \mathbb{R}/_M)}{\# M_{tors}}\right)$$

$$\chi(M) := \tilde{\chi}(M) - \tilde{\chi}(R) \cdot \mathrm{Rang}(M)$$, where R obtains the standard Haar measure on $(R \otimes_{\mathbb{Z}} \mathbb{R})$.

$\tilde{K}_0(R)$ should be the Grothendieck group generated by the finitely generated R-modules with volume form on $M \otimes_{\mathbb{Z}} \mathbb{R}$, (M, vol) .

The relations are given by the exact sequences

$$0 \to M_1 \to M \to M_2 \to 0 \quad ,$$

such that $\lambda(M \otimes_{\mathbb{Z}} \mathbb{R}) \cong \lambda(M_1 \otimes_{\mathbb{Z}} \mathbb{R}) \otimes \lambda(M_2 \otimes_{\mathbb{Z}} \mathbb{R})$ as hermitian line bundles. (under the canonical map)

It is easy to check, that one has a mapping

$$\left. \begin{array}{l} \chi : \widetilde{K}_o(R) \longrightarrow \mathbb{R} \\ \quad (M,\mathrm{vol}) \longmapsto \chi(M) \end{array} \right\}$$

Therefore we define

Definition 2: If $\mathcal{L}$ is a hermitian line bundle on X . Then we pose

$$\chi(\mathcal{L}) := \chi(\mathbb{R}\Gamma(X,\mathcal{L}))$$

The main result of this section is

Theorem 1: (Riemann-Roch) One has $\chi(\mathcal{L}) = \frac{1}{2} < \mathcal{L}, \mathcal{L} - \omega_X > + \chi(\mathcal{O}_X)$, where $\omega_X = \omega_{X/B}$ is the relative dualizing sheaf of X over B.

Proof. We proceed as in [F]

i) The formula holds for $\mathcal{L} = \mathcal{O}_X$. Suppose, the formula is true for $\mathcal{L} = \mathcal{O}_X(D)$. We have to show, it remains true, if one adds an arbitrary divisor D_o .

ii) $D_o = \alpha_v F_v$, $v \in S_\infty$, $\alpha_v \in \mathbb{R}$. We obtain (writing $\chi(D)$ instead of $\chi(\mathcal{O}_X(D))$)

$$\chi(D + \alpha_v F_v) - \chi(D)$$
$$= \alpha_v((h^o(D) - h^1(D))|_{F_v})$$
$$= \alpha_v(\deg(D)|_{F_v} + 1-g$$
$$= <D, \alpha_v F_v> - \frac{1}{2} <F_v, \omega_X> \qquad \text{okay.}$$

iii) Suppose, $D_o = C$ is an irreducible component of a fibre. We have to compute $\chi(D+C) - \chi(D)$. But we have the exact sequence

$$0 \to \mathcal{O}_X(D) \to \mathcal{O}_X(D+C) \to \mathcal{O}_X(D+C)/\mathcal{O}_X(D) \to 0$$

We obtain the following equation in $\tilde{K}_o(R)$:

$$\left.\begin{aligned} &\mathbb{R}\Gamma(\mathcal{O}_X(D)) + \mathbb{R}\Gamma(\mathcal{O}_X(D+C)/\mathcal{O}_X(D)) \\ &= \mathbb{R}\Gamma(\mathcal{O}_X(D+C))) \end{aligned}\right\}$$

Using property iii) of the volume forms, defined in §3. Therefore

$$\chi(D+C) - \chi(C) = \chi(\mathcal{O}_X(D+C)/\mathcal{O}_X(D))$$
$$= \log \#(\mathcal{O}_X(D+C)/\mathcal{O}_X(D))$$
$$= \log(q_v)(\deg(C+D)|_C + 1-g_C)$$
$$= <C+D, C> - \frac{1}{2} <C, C+\omega_X> ,$$

because $\quad g_C = 1 + \frac{1}{2\log(q_v)} <C, C+\omega_X>$

using the adjunction formula.

iv) $D_o = s(B)$, where $s:B \to X$ is a section for $\pi:X \to B$, given by $P \in X_\eta(K)$

We have

$$\chi(D+s(B)) - \chi(D)$$

$$= \chi(\mathcal{O}_X(D+s(B))/\mathcal{O}_X(D))$$

$$= \chi(s^*(\mathcal{O}_X(D+s(B))))$$

$$= \langle D+s(B) , s(B)\rangle$$

But we have the following

<u>Lemma 1:</u> One has an isomorphism

$$s^*(\omega_X \otimes \mathcal{O}_X(P)) \xrightarrow{\sim} R$$

Proof: One can define a map using residues. The surjectivity of the map can be tested locally.

Therefore we obtain

$$\langle D + s(B) , s(B)\rangle$$

$$= \langle D-\omega_X , s(B)\rangle$$

$$\left.\begin{aligned} &= \frac{1}{2} \langle D+s(B) , D+s(B) -\omega_X\rangle \\ &- \frac{1}{2} \langle D, D-\omega_X\rangle \end{aligned}\right\}$$

q.e.d.

§ 5 The Hodge index theorem

We have the same notations as in §4.

Theorem 1 (Hodge index theorem)

Denote V_v for $v \in S$, the set of places of R, the set of Arakelov divisors, which are generated by irreducible components of the fibre F_v. Then the following holds

1) The intersections pairing $<,>$ is negativ semidefinit on V_v. The same is true for $\oplus V_v$.

2) Suppose $D \in \widetilde{\mathrm{Div}}(X)$ and $D \perp V_v$ for all $v \in S$. Then $\mathcal{O}_X(D)|_{X_\eta} \in \mathrm{Jac}(X_\eta)(K)$.

$$\text{One has:}\quad \left.\begin{aligned} <D,D> &= -2(K:\mathbb{Q}) \\ &\cdot \text{Néron Tate height } (\mathcal{O}_X(D)|_{X_\eta}) \end{aligned}\right\}$$

(as an element of $\mathrm{Jac}(X_\eta)(K)$

3) The signature of $<,>$ on the group $\widetilde{\mathrm{Div}}(X)/\{(f)^\sim | f \in K(x)\}$ is sign $(< >) = (+,-,\dots,-)$ and the number of -signs is $\#(-) = \sum_{v \in S} ((\#\text{ of components of } F_v)-1) + \mathrm{Rang}\ \mathrm{Jac}(X_\eta)(K)$.

For 1.) one can proceed exactly as in the classical situation. The reader can consult [F] if necessary.

3.) follows from 1.) and 2.) It remains to show 2.):

Because $<D,F_v> = 0$, we can conclude: $\deg(\mathcal{O}_X(D)|_{X_\eta})=0$.

Therefore we obtain a class

$$(\mathcal{O}_X(D)|_{X_\eta}) \in \mathrm{Jac}(X_\eta)(K) .$$

The line bundles of degree $(g-1)$ on X give a scheme $\mathrm{Pic}_{g-1}(X/B)$ over B, locally of finite type over B.

There exists an open subscheme

$$P \subset \mathrm{Pic}_{g-1}(X/B) ,$$

where we have removed all components in F_v , $v \in S$, except the one, which contains a fixed line bundle $\mathcal{O}(E)$.

Then P will be of finite type over B . Consider again our D above ,

$$D \perp V_v \quad \forall \quad v \in S$$

Then $\mathcal{O}_X(E+nD)$ will define a point in P for all $n \in \mathbb{Z}$. We consider $\theta \subset P$ as the closure of the standard theta-divisor on $P_\eta = \mathrm{Pic}_{g-1}(X_\eta)/K$. We have seen in §3 , that if $\mathcal{L}$ denotes the universal line bundle over P , we have the isomorphism

$$\lambda(R\pi_*(\mathcal{L})) = \mathcal{O}_P(-\theta)$$

and this is even an isomorphism of hermitian line bundles on P , as we have seen in §3 .

Now, the class of $\mathcal{O}_X(E+nD)$ defines a rational section $B \xhookrightarrow{s} P$.

We obtain the following diagram

$$\begin{array}{ccccc} X \times_B (s) & \rightarrow & X \times_B P & \subset & X \times_B \mathrm{Pic}_{g-1}(X/B) \\ \pi \downarrow & & \pi \downarrow & & \pi \downarrow \\ B & \xhookrightarrow{s} & P & \subset & \mathrm{Pic}_{g-1}(X/B \end{array}$$

Using base change again, we obtain

$$s^*(\mathcal{O}_P(-\theta)) = s^* \lambda \mathbb{R}\pi_*(\mathcal{L})$$

$$= \lambda(\mathbb{R}\pi_*(s^*(\mathcal{L}))$$

$$= \lambda \mathbb{R}\,\pi_*(\mathcal{O}_X(E+nD))$$

These are isomrophisms as hermitian line bundles on B, because the isomorphismsare given canonically and we have seen in §3 that these canonical isomorphismsinduce isometriesfor the Riemann surfaces X_v, $v \in S_\infty$.

We therefore can conclude:

$$\deg(s^*\mathcal{O}_P(-\theta))$$

$$= \deg(\lambda\,\mathbb{R}\pi_*\mathcal{O}_X(E+nD))$$

$$= \chi(\mathcal{O}_X(E+nD)) \quad .$$

Using the Riemann-Roch theorem on the one hand we have

$$\chi(\mathcal{O}_X(E+nD))$$

$$= \chi(\mathcal{O}_X) + \frac{1}{2} <E+nD, E+nD-\omega_X>$$

$$= \frac{n^2}{2} <D,D> + \text{terms, only linear in } n .$$

On the other hand, using the results of part II , Heights, we immediately obtain the equality

$$\deg(s^{*}\,\Theta_{P}(-\theta)) = (K:\mathbb{Q}) \times \text{ logarithmic height } (E+nD)$$

Using the relation of logarithmic height and Néron-Tate height, the result follows. q.e.d.

REFERENCES:

[A1] S. Arakelov: Families of curves whith fixed degeneracies, Izv. Akad. Nauk. 35, 1971, 1269-1293.

[A2] S. Arakelov: An intersection theory for divisors on an arithmetic surface, Izv. Akad. Nauk 38, 1974, 1179-1192.

[A3] S. Arakelov: Theory of Intersections on the Arithmetic surface, Proc. Int. Congress Vancouver, 1974, 405-408.

[F] G. Faltings: Calculus on arithmetic surfaces, Annals of Math., 1984, to appear.

[F1] G. Faltings: Properties of Arakelov's Intersection product. SLN 997, p. 138-146.

[G-H] Ph. Griffith, J. Harris: Principles of algebraic Geometry, New York, 1978.

[Q] D. Quillen: Determinants of $\bar{\partial}$-operators, Vortrag auf der Bonner Arbeitstagung 1982.

[H] P. Hriljac: Heights and Arakelov's intersection theory, Am. J. Math. 107, 23-38.

APPENDIX

New Developments in Diophantine and Arithmetic Algebraic Geometry

Gisbert Wüstholz

Contents:

1 Introduction

Since this book has appeared the subject to which it is devoted has developed enormously. Faltings' work had significant influence on the theory of diophantine inequalities, diophantine geometry, arithmetic intersection theory, the theory of ℓ-adic representations and on the theory of Riemann surfaces and string theory. In this appendix we shall give an overview on parts of developments without making an attempt to be complete.

Since Faltings proved the Mordell conjecture three entirely different methods have been developed for proving the finiteness of rational points on curves of genus at least 2. The first approach by D.W. Masser and G. Wüstholz bases on methods in transcendence theory which were introduced into the field by these authors and have their roots in the work of A. Baker on linear forms in logarithms. The second entirely different new method was created by P. Vojta and roughly speaking is a substantial generalization of the theory of rational approximations to algebraic numbers which goes back to Thue, Siegel, Schneider, Dyson and Roth. The third alternative approach was given by G. Faltings and goes beyond the previous work. He proves that on a subvariety of an abelian variety over a number field there are only finitely many rational points unless this subvariety contains a translate of a proper and nontrivial abelian subvariety. In this case clearly such a subvariety has in general infinitely many rational points. Faltings' work heavily bases on Vojta's achievements. Finally E. Bombieri gave a beautiful rather elementary variant of Vojta's proof. Vojta uses in his proof in a significant way the intersection theory on arithmetical surfaces which was introduced by Arakelov and further developed by Faltings and others (see Chapter VII of the book). Also this theory has made further progress and has been documented in Lang's book [L1]. Basically Vojta's proof is the first and only significant application of this theory which lay the foundations for the higher dimensional arithmetic intersection theory. This generalization to higher dimension was developed by H. Gillet and Ch. Soulé with important contributions made by Bismut. In fact Vojta's proof uses it only in special cases and here in particular an asymptotic version of the Grothendieck-Hirzebruch-Riemann-Roch theorem which was proved by these authors. More generally they published a conjectural arithmetical version of the exact Grothendieck-Hirzebruch-Riemann-Roch theorem. A complete answer for this conjecture was then given by G. Faltings in [F1].

The difficult Arakelov theory then disappears in Faltings' proof of the result on subvarieties of abelian varieties mentioned above and is replaced by a refined version of Siegel's Lemma and a new height theory in projective space. In Bombieri's version of Vojta's proof finally it turns out that the classical height theory suffices for obtaining a proof of Mordell's conjecture.

Faltings' original proof of the Mordell conjecture makes heavily use of the compactification of the moduli space of principally polarized abelian varieties. At the time Faltings had only ad hoc methods to his disposal. This was the starting point for intensive investigations mainly by C.L. Chai and G. Faltings. In their book [F-Ch] they construct for example the compactification of the moduli stack of principally polarized abelian varieties over the integers which can be used for example to replace the ad hoc theory of heights on ablian varieties in our book.

The theory of ℓ-adic representations attached to abelian varieties was further developed by J-P. Serre. He studied for example how large the image of the Galois group can be compared with the general linear group over the ℓ-adic integers or over the field $\mathbb{F}_\ell$. A short report on his work is given in section 11.

2 The Transcendental Approach

In the past two decades transcendence theory has developed very powerful tools in order to deal with diophantine questions. One striking example was for instance Shafarevich's theorem on the finiteness of the number of isogeny classes of elliptic curves over a number field with good reduction outside a finite set of finite places. Here the problem of bounding this number was reduced to the famous Mordell equation $y^2 = x^3 + k$ for which effective upper bounds for the height of the solutions in terms of k were given using the theory of linear forms in logarithms. Another example is the proof of Tate's conjecture for elliptic curves over the field of rational numbers given by D. and G. Chudnovsky [Ch]. Recently an entirely new method was found by D.W. Masser and G. Wüstholz in order to prove Tate's conjecture for elliptic curves in general. This approach bases on linear forms in elliptic logarithms. They prove an estimate for isogenies between elliptic curves from which it is also not very difficult to deduce Shafarevich's theorem. In order to state the result let E be the elliptic curve given by

$$y^2 = 4x^3 - g_2 x - g_3$$

where g_2, g_3 are algebraic numbers and let the height of the elliptic curve be given by

$$h(E) = \max(1, h(g_2), h(g_3))$$

and where $h(g_i)$ is the absolute Weil height of g_i $(i = 1, 2)$.

2.1 Isogeny Theorem (for elliptic curves [MW1]) *Given a positive integer d, there exists a constant c, depending effectively on d, with the following property.*

Let K be a number field of degree at most d, and let E be an elliptic curve defined over K. Suppose that E' is an elliptic curve over K isogenous to E. Then there is an isogeny ϕ between E and E' with

$$\deg \phi \leq c \cdot h(E)^4 \ .$$

We note that the constant c depends only on the degree and not on the discriminant of K. The idea of the proof is very simple to explain. Let Ω and Ω' be the period lattice for E and E' respectively. Then an isogeny ϕ induces a homomorphism ϕ_* from Ω to Ω'. Hence one obtains an algebraic number $\alpha = \alpha(\phi)$ with

$$\alpha\Omega \subseteq \Omega' \ .$$

This gives a set of dependence relations

$$\alpha\omega_i = m_{i1}\omega_1' + m_{i2}\omega_2' \quad (i = 1, 2)$$

where ω_1, ω_2 and ω_1', ω_2' are a basis for Ω and Ω' respectively. By transcendence methods one obtains new relations with bounds for m_{ij} $(i, j = 1, 2)$ depending only on d and $h(E)$. The new relations give then a new isogeny ϕ_0 with bounds on the degree as stated in the theorem.

As an application an effective bound for the number of K-isomorphism classes of elliptic curves defined over K which are K-isomorphic to E is obtained (see [MW2]). This number is at most equal to $ch(E)^8$.

The basic idea for the proof of the Isogeny Theorem for elliptic curves generalizes of course to higher dimensional abelian varieties. However, many technical problems arise. These were solved in [MW3] and an abelian analogue of the elliptic isogeny theorem is proved in [MW4]. The heart of the proof of the abelian isogeny theorem is an effective version of the analytic subgroup theorem in [W1].

In order to state it let A be an abelian variety over K, L a positive element in $(\text{Pic } A)(K)$. Then L determines a hermitian form $H(z, w)$ on $(\text{Lie } A) \otimes \mathbb{C}$. If we let E be the imaginary part of H and write $A(\mathbb{C})$ as V/Λ for a complex vectorspace and a lattice $\Lambda \subseteq V$ then E is integer valued on Λ and non-degenerate. The choice of a symplectic basis $\gamma_1, \ldots, \gamma_g, \delta_1, \ldots, \delta_g$ of Λ determines a point $\tau = (\tau_{ij})$ in the Siegel upper half plane $\mathbb{H}_g$ by expressing the γ's in terms of the δ's:

$$\gamma_j = \sum_{i=1}^{g} \tau_{ij}\delta_j \ .$$

In this basis H takes then the form

$$H(z, w) = {}^t z (Im\tau)^{-1}\bar{w} \ .$$

If we put $\varepsilon = (E(\gamma_i, \delta_j))$ then ε is a diagonal matrix and the period matrix for A is given by (ε, τ). The symplectic group $\Gamma = Sp_{2g}(\mathbb{Z})$ operates on $\mathbb{H}_g$ and its quotient is the moduli space of principally polarized abelian varieties. Since L does not necessarily belong to the class of principal polarization we have to use level structures. This means that we have to replace $\Gamma \backslash \mathbb{H}_g$ by a finite covering which depends on the type of polarization. Next we fix a fundamental domain in $\mathbb{H}_g$ for the normal subgroup Γ' of Γ which corresponds to this covering and choose A in its isomorphism class such that τ is in this fundamental domain. We then fix a basis of Lie A. This can be done in a natural way depending on L and the choice of the symplectic basis above. It enables us to define the height of a linear subspace X of Lie A as the Weil height of the point in the Grassmann variety determined by X. The following result on diophantine approximations is basic for all further applications.

2.2 Main Theorem. *Let $0 \neq \omega \in (X \otimes \mathbb{C}) \cap \Lambda$. Then there exists an abelian subvariety H over K of A such that $\omega \in (\text{Lie } H) \otimes \mathbb{C} \subseteq X \otimes \mathbb{C}$. Furthermore, if $h \geq 1$ is a real number such that $h_F(A) \leq h$, then*

$$\deg H \leq c_0 \left(h + H(\omega, \omega) + h(X)\right)^{c_1} .$$

Here the constants c_0 and c_1 are effective. The constant c_0 depends only on the degree d of K, the degree δ of the polarization L and the dimension g of A; c_1 depends only on g.

For ω as in the Main Theorem we let H_ω be the smallest abelian subvariety H such that $\omega \in (\text{Lie } H) \otimes \mathbb{C}$. Then it is easy to deduce from the Main Theorem the following result.

2.3 Effective Algebraic Subgroup Theorem ([MW3]). *There exist effective constants c_0, c_1 such that*

$$\deg H_\omega \leq c_0 \left(h + H(\omega, \omega)\right)^{c_1} .$$

Here c_0 depends only on d, δ and g and c_1 depends only on g.

By geometry of numbers one can always find a basis $\omega_1, \ldots, \omega_{2n}$ for the period lattice such that $H(\omega_j, \omega_j)$ is bounded in terms of h. We shall later apply the algebraic subgroup theorem in the case when ω is such a basis element. This means that $H(\omega, \omega)$ in the statement of the theorem can be eliminated.

We shall now briefly indicate the main ideas for the proof of the Main Theorem. For this we choose a sufficiently large integer D depending explicitly on $h_F(A), H(\omega,\omega)$ and $h(X)$. The line bundle L carries in a natural way a hermitian metric given by the hermitian form above. In the first step of the proof a non-trivial integral section σ in $\Gamma(A, L^D)$ is constructed such that σ vanishes at the origin to a high order T with respect to X. The parameters D and T are linked via Riemann-Roch. This theorem gives the proportion between the number of condition and the degree of freedom of the system of equations in question. The norms of σ at the infinite places can be bounded in terms of $h_F(A)$, $H(\omega,\omega)$ and $h(X)$ by Siegel's Lemma. Using the Schwarz Lemma it is shown in the next step that σ vanishes to order $[T/2]$ with respect to S at those ℓ-torsion points of A for sufficiently large prime ℓ which are in X. The set of such points is non-empty since $0 \neq w \in (X \otimes \mathbb{C}) \cap \Lambda$ by hypothesis. The prime number ℓ is chosen sufficiently big such that the number of condition imposed on σ by these vanishing properties is too large unless some degeneration happens. This is shown using the multiplicity estimates given in[W2], and there the degeneration is also described explicitly. It is given by an algebraic subgroup H of A with the properties stated in the Main Theorem, except that it does not follow from the multiplicity estimates that ω is in (Lie H)$\otimes\mathbb{C}$. In order to achieve this a rather technical inductive procedure has to be used.

The Algebraic Subgroup Theorem implies now the following result on isogenies.

2.4 Isogeny Theorem ([MW4]). *There exist effective constants $c, \kappa > 0$ with the following property. Let $h \geq 1$ be a real number and A, B be principally polarized abelian varieties over K. Then, if A and B are isogenous and* $\min(h_F(A), h_F(B)) \leq h$, *there exists an isogeny ϕ from A to B over K such that*

$$\deg \phi \leq ch^{\kappa} .$$

The constant c depends only on the dimension g of A and d, κ depends only on g.

For the proof one starts with a minimal isogeny ψ from A to B and in the same way as for elliptic curves one obtains a set of period relations between the periods of A and the periods of B. More precisely: if $\omega_1, \ldots, \omega_{2g}$ and $\omega'_1, \ldots, \omega'_{2g}$ denote a basis for the period lattices of A and B one gets a relation

$$d\phi(\omega) = m_1\omega'_1 + \ldots m_{2g}\omega'_{2g}$$

where $m_1, \ldots, m_{2g}$ are integers, $d\phi$ is the tangent map and ω is one of the ω_j. We put $G = A \times B^{2g}$ and define the subspace X of Lie G by the equation

$$d\phi(z) = m_1 z_1 + \ldots + m_{2g} z_{2g}$$

where z is a coordinate vector in Lie A and the z_j are independent coordinate vectors in Lie B for $j = 1, \ldots, 2g$. Then $\eta = (\omega, \omega_1, \ldots, \omega_{2g})$ is in X and nonzero. By the Effective Algebraic Subgroup Theorem we can bound the degree of $H_\eta \subseteq G$ and one shows that H_η leads to an isogeny from an abelian subvariety $B' \subseteq B$ to an abelian subvariety $A' \subseteq A$. Furthermore all the degrees involved can be bounded in terms of the degree of H. In the case that A is simple we are finished. Otherwise we have to use induction which again is somewhat technical.

Clearly the isogeny theorem gives immediately the finiteness of the number of isomorphism classes of principally polarized abelian varieties within one isogeny class. Using the results of Tate also Tate's conjecture follows directly. As a consequence one gets the semi-simplicity of the Tate module which implies the finiteness of the number of isogeny classes as in chapter V of the book. Then all the other finiteness theorems follow as in Faltings' original proof. The isogeny theorem leads also to bounds for the heights of abelian varieties within one isogeny class. On the other hand it is not known whether a bound for the absolute value of the difference of the heights of two isogenous abelian varieties over K yields an estimate on the degree of a minimal isogeny between them.

In order to give some further applications let first End A be the ring of endomorphisms of A. This is a free group over the integers of finite rank. We fix a positive element ζ in the Néron-Severi group $NS(A)$ of A and obtain an isogeny $\wedge(\zeta)$ from A to the abelian variety $\hat{A}$ dual to A such that (by abuse of notation)

$$\wedge(\zeta)^{-1} \circ \wedge(\zeta) = \deg \zeta \cdot id_A$$

where $\deg \zeta$ denotes the degree of $\wedge(\zeta)$. The 'Rosati involution' on End A is then defined by

$$\alpha \mapsto \alpha^t = \wedge(\zeta)^{-1} \circ \hat{\alpha} \circ \wedge(\zeta)$$

and using this involution one defines a bilinear form on End A. It is given by

$$(\alpha, \beta) \mapsto \langle \alpha, \beta \rangle = Tr(\alpha \circ \beta^t) \ .$$

The bilinear form $\langle , \rangle$ takes integer values and is positive definite on End A. We use it to define the discriminant of End A with respect to ζ. For this let $\varepsilon_1, \ldots, \varepsilon_m$ be a basis for End A. Then we put

$$\mathcal{D}_\zeta := \det \left((\langle \varepsilon_i, \varepsilon_j \rangle)_{ij} \right) \ .$$

The integer $\mathcal{D}_\zeta$ is positive and called the discriminant of End A with respect to ζ. Note that in general it is different from the algebra discriminant $\mathcal{D}$.

2.5 Discriminant Theorem ([MW5]). *Let d, δ, h and n be positive integers. There exists a constant c depending only on d, δ and n, and a constant κ depending* *only on n with the following propery. Let A be an abelian variety of dimension at most n defined over a number field of degree at most d with Faltings height at most h and let ζ a positive element in NS(A) of degree at most δ. Then we have*

$$\mathcal{D}_\zeta \leq ch^\kappa .$$

The constants c and κ are effectively computable.

The main ingredient for the proof of the discriminant theorem is the algebraic subgroup theorem. First using this theorem one shows that every endmorphism ϕ in End A can be written as a linear combination

$$\phi = v_1\phi_1 + \ldots + v_m\phi_m$$

of endomorphisms $\phi_j \in$ End A with rational coefficients such that, if one defines $||\phi||^2 = \langle\phi,\phi\rangle$,

$$||\phi_j||^2 \leq c'h^{\kappa'}$$

with c' and κ' as in the theorem. The second step is an easy calculation using Hadamard's inequality. Namely among the elements ϕ_j one takes a set of linearly independent endomorphisms which generate a sublattice of End A. The discriminant for this sublattice is an upper bound for the discriminant of End A and an appeal to Hadamard in the case of the sublattice yields the result.

For a sketch of the first step we assume that A is simple in order to simplify the argument. Let ϕ be an endomorphism of A and choose a basis $\omega_1,\ldots,\omega_{2n}$ for the lattice of periods of A. Then for any period vector ω in the lattice we get as before a period relation

$$d\phi(\omega) = m_1\omega_1 + \ldots + m_{2n}\omega_{2n}$$

with integers $m_1,\ldots,m_{2n}$. We define the algebraic subgroup $\Gamma \subset A^{2n} \times A$ by

$$\phi(P) = m_1P_1 + \ldots + m_{2n}P_{2n}$$

where $(P_1,\ldots,P_{2n})$ is in A^{2n} and P in A. The algebraic subgroup contains the 1-parameter subgroup determined by $\eta = (\omega_1,\ldots,\omega_{2n},\omega)$. The algebraic subgroup theorem gives then an algebraic subgroup $H_\eta \subseteq \Gamma$. We let $H \subseteq \Gamma$ be the orthogonal complement with respect to ζ in H_η of the connected component of the neutral element of $H_\eta \cap (A^{2n} \times 0)$. Then H is isogenous to A and can be parametrized by

$$(\phi_1(P),\ldots,\phi_{2n}(P),m_0P)$$

where $\phi_1,\ldots,\phi_{2n}$ are in End A and m_0 is a positive integer. Furthermore the relation above gives

$$m_0\phi(P) = (m_1\phi_1 + \ldots + m_{2n}\phi_{2n})(P)$$

identically in P. Hence we get

$$m_0\phi = m_1\phi_1 + \ldots + m_{2n}\phi_{2n} \quad .$$

Since the quantities $||\phi_j||$ can be bounded by the degree of H which itself is bounded in terms of the degree of H_η the desired inequalities are established.

The bound for the discriminant $\mathcal{D}_\zeta$ in the discriminant theorem depends on the degree of the polarization. An obvious question then arises whether the minimal degree of a polarization defined over K can be bounded in terms of the degree d of K, n and the Faltings height $h_F(A)$. A partial answer to this question is given by the following result.

2.6 Polarization Theorem ([MW6]). *Let $n \geq 1, d \geq 1$ and $h \geq 1$ be real numbers. There exists a constant c depending only on n and d, and a constant λ depending only on n, with the following property. Suppose A is an abelian variety of dimension at most n defined over a number field of degree at most d with Faltings height bounded by h and assume that End $A = \mathbb{Z}$. Then A has a polarization of degree at most ch^λ.*

The isogeny theorem can be applied to Galois representations attached to elliptic curves. Let K be a number field, $\bar{K}$ its algebraic closure and $\pi = \mathrm{Gal}(\bar{K}/K)$. If E is an elliptic curve defined over K, then π acts on $E(\bar{K})$ and in particular on the group E_ℓ of points of order dividing ℓ. When ℓ is a prime number then E_ℓ is a vector space over $\mathbb{F}_\ell = \mathbb{Z}/\ell\mathbb{Z}$. We obtain in this way a representation $\rho_\ell : \pi \rightarrow \mathrm{GL}(E_\ell)$. A fundamental result of Serre ([S1], [S2]) says that if E has no complex multiplication over $\bar{K}$ then there exists a constant $\ell_0 > 0$ such that $\rho_\ell(\pi) = \mathrm{GL}(E_\ell)$ for all $\ell > \ell_0$.

Up to now it seems that no general estimate for ℓ_0 has been written down. Serre gives a number of examples and some results for special classes of elliptic curves. For example he obtains in [S2] a simple estimate when $K = \mathbb{Q}$ and E is semistable. In his later paper [S3] he removes the semistability condition by assuming the Generalized Riemann Hypothesis. In [MW7] a general estimate for ℓ_0 is given. For this let E be without complex multiplication and $h \geq 1$ a real number such that $h_F(E) \leq h$.

2.7 Theorem. *There are absolute positive constants c, γ with the following properties*

(i) $\rho_\ell(\pi) \supseteq \mathrm{SL}(E_\ell)$ if $\ell > c(\max(d, h))^\gamma$,

(ii) if further ℓ does not divide the discriminant of K then $\rho_\ell(\pi) = \mathrm{GL}(E_\ell)$.

3 Vojta's Approach

In the last hundred years the theory of diophantine approximations developed very powerful tools in order to show that certain diophantine inequalities and, as a consequence, a large class of diophantine equations have only finitely many rational or integer solutions. One central subject was the approximation of algebraic numbers by rationals and the theory culminated in the famous theorem of Roth [Ro] who showed that for any positive real number δ and for any given algebraic irrationality α the inequality

$$\left|\alpha - \frac{p}{q}\right| < q^{-2+\delta}$$

has only finitely many solutions in rational numbers p/q such that $(p,q) = 1$ and $q > 0$.

Very surprisingly in a fundamental paper Vojta [V1] succeeded in 1989 to find a new proof of Mordell's conjecture for function fields along the lines of the proof of Roth's theorem. Formally both proofs are very similar. But a deep insight was necessary to make the method work in a very different situation. It was then not surprising that recently Vojta [V2] was able to further extend his methods and to make the proofs also work in the arithmetical case and to obtain a new proof of Faltings' theorem. In order to indicate the main ideas of his proof we shall first give a short description of the proof of Roth's theorem. Then we shall describe Vojta's proof of Mordell's conjecture in the function field case and in Section 7 we shall discuss the main new ingredients for obtaining the arithmetic analog. One of the basic concepts in the proof of Roth's theorem is the index of a polynomial in several variables. More generally let $\varphi(t_1,\ldots,t_m)$ be a power series in $t_1,\ldots,t_m$ with complex coefficients and $d_1,\ldots,d_m$ be any positive integers. If $\xi_1,\ldots,\xi_m$ are complex numbers such that φ is holomorphic at $\xi = (\xi_1,\ldots,\xi_m)$ then the index $i(\varphi,\xi)$ of φ at ξ with respect to $d_1,\ldots,d_m$ is defined to be the least value of $i_1/d_1 + \ldots + i_m/d_m$ such that

$$(\partial/\partial t_1)^{i_1} \ldots (\partial/\partial t_m)^{i_m} \varphi(\xi_1,\ldots,\xi_m) \neq 0 \ .$$

If φ vanishes identically then we put $i(\varphi,\xi) = +\infty$. This definition extends clearly to arbitrary fields of characteristics 0. We fix now $\varepsilon > 0$ sufficiently small with respect to δ and an integer m sufficiently large depending on $1/\varepsilon$. Then the first step of the proof of Roth's theorem consists in constructing a polynomial $P \neq 0$ in

$t_1, \ldots, t_m$ with intger coefficients such that

$$\begin{aligned} &\deg_{t_i} R \leq d_i \quad (1 \leq i \leq m) , \\ &i\,(P, (\alpha, \ldots, \alpha)) \geq (m/2)(1 - \varepsilon) , \\ &h(P) << d_1 + \ldots + d_m \quad . \end{aligned}$$

Here $h(P)$ is the logarithmic Weil height of P. Such a polynomial is obtained by Siegel's Lemma which one derives small solutions for systems of equations in more variables than equations.

In the second step one assumes that the diophantine inequality in question has infinitely many solutions. Then it is possible to choose solutions p_i/q_i $(i = 1, \ldots, m)$ such that

$$d_1 \log q_1 \leq d_i \log q_i \leq (1 + \varepsilon) d_1 \log q_1 \quad .$$

This fixes the quotients d_i/d_1. Expanding P around $(\alpha, \ldots, \alpha)$ and using the fact that $|\alpha - p_i/q_i| < q_i^{-2-\delta}$ $(i = 1, \ldots, m)$ it is easy to prove that

$$i\,(P, (p_1/q_1, \ldots, p_m/q_m)) \geq \varepsilon m \quad .$$

The last step then consists of producing an upper bound for this index. There are two methods for obtaining such an upper bound. Roth uses the so called Roth Lemma. Namely under the additional condition that

$$d_i/d_{i+1} \geq \omega \quad (i = 1, \ldots, m - 1)$$

for a sufficiently large positive number ω depending on ε one gets

$$i\,(P, (p_1/q_1, \ldots, p_m/q_m)) \leq \varepsilon \quad .$$

This extra condition can easily be satisfied by the choice of the p_i/q_i. Comparing finally the upper and lower bound for the index one derives a contradiction and this establishes the finiteness of the number of solutions of the diophantine inequality in question.

Technically Vojta's method follows more the approach by Siegel and Dyson who used two variables $(m = 2)$ and so not obtaining the best possible exponent $2 + \delta$ instead of large m as in Roth's approach. For proving the finiteness of rational points this is completely sufficient. However, it would be interesting to try to make the method also work for large m such that possibly good bounds for the number of solutions can be obtained. For this one would have to extend the generalization of Dyson's Lemma to more than two variables which was obtained by E. Viehweg and H. Esnault to a more general geometric situation. Any progress in this direction would be of considerable interest.

Let now B be a smooth connected projective curve over an algebraically closed field of characteristic zero or the spectrum of the ring of integers $\mathcal{O}_K$ of an algebraic number field K. Then let $\pi : X \to B$ be a regular semi-stable family of curves over B with smooth generic fibre C of genus $g > 1$ and K_C the canonical divisor on C. Then $\mathrm{Num}(C \times C) \otimes \mathbb{R}$ is a finite dimensional vector space and the intersection pairing is a nondegenerate bilinear form. Denote by p_i the projection from $C \times C$ onto the i-th factor for $i = 1, 2$. Then the elements $F_i = p_i^* K_C/(2g-2)$ for $i = 1, 2$ and the diagonal Δ generate a subspace E on which the intersection matrix in terms of the basis $F_1, F_2, \Delta' = \Delta - F_1 - F_2$ takes the form

$$\begin{pmatrix} 0 & 1 & 0 \\ 1 & 0 & 0 \\ 0 & 0 & -2g \end{pmatrix} .$$

One considers now the element $V_r = \Delta' + a_1 F_1 + a_2 F_2$ in E where $a_i = a_i(r)$ are rational numbers which depend on a parameter r for $i = 1, 2$ such that

$$V_r^2 = 2\delta > 0$$

is independent on r. If r is sufficiently large then one shows that V_r is ample.

The next step is to extend everything to the base scheme B. Let W be the desingularization of $X \times_B X$ and $q : W \to B$ be the structural morphism. We fix a fibre F of q. Taking closures in W we extend Δ', F_1, F_2 to B and define for rational numbers b the divisor

$$\tilde{V}_r = \Delta' + a_1 F_1 + a_2 F_2 + bF \ .$$

It is shown that for sufficiently large b and sufficiently large integers d depending on b and r the divisor $d \cdot \tilde{V}_r$ is linearly equivalent to an effective divisor V. For the rest of this section we restrict ourselves to the function field case where B is a geometric curve. Then we write $\mathcal{L}_d = \mathcal{O}(d\tilde{V}_r)$ and obtain by the Grothendieck-Riemann-Roch theorem that

$$\sum_{i=0}^{2} (-1)^i \deg R^i q_* \mathcal{L}_d = \frac{d^3}{6} \tilde{V}_r^3 + O(d^2) \ .$$

Then one shows that $\deg R^1 q_* \mathcal{L}_d \geq 0$ and that $R^2 q_* \mathcal{L}_d = 0$ for d sufficiently large. Thus one gets

$$\deg q_* \mathcal{L}_d \geq c_1 d^3$$

for some positive constant c_1. Also

$$\mathrm{rank} \ q_* \mathcal{L}_d \leq c_2 d^2$$

for some further constant c_2. An appeal to Riemann-Roch yields then a global section of $q_* \mathcal{L}_d$ and therefore also of $\mathcal{L}_d$.

Let now P_1, P_2 be two different rational points on C and E_1, E_2 the corresponding sections of π. We take the intersection product $E = (p_1^* E_1 \cdot p_2^* E_2)$ which is just (P_1, P_2) on the generic fibre $C \times C$. The next step is to compute $(E \cdot V)$. This amounts to calculate the numbers $(E \cdot \Delta), (E.F_1), (E \cdot F_2)$ and $(E \cdot F)$. Clearly $(E \cdot F_i) = \left(E_i \cdot \frac{\omega_{X/B}}{2g-2}\right) = h_i$, $(E \cdot F) = 1$ and using a well-known result of Mumford [M] one finds that

$$(E \cdot \Delta) = (E_1 \cdot E_2) \leq h_1 + h_2 - 2\gamma\sqrt{h_1 h_2} + c$$

for some constant $\gamma > 0$ such that $\gamma + \delta < g$ and some constant c, provided h_1 and h_2 are sufficiently large. They are chosen in such a way that r is approximately equal to h_2/h_1. Then an easy calculation yields

$$(E \cdot V) = (E \cdot d\tilde{V}_r) \leq -\eta d\, h_1 \cdot \sqrt{r}$$

for some positive constant η. This means that E is contained in V. If we write V as the divisor of a section σ this implies that σ vanishes in E. We pull σ back to $C \times C$, trivialize it in a neighbourhood of (P_1, P_2) and obtain a power series $\varphi(t_1, t_2)$ in fixed chosen uniformizing parameters t_1, t_2 around P_1, P_2. The next rather technical step is to sharpen this vanishing property and to show that the index $i(\sigma, E) := i(\varphi, (0,0))$ is bounded asymptotically below by

$$\frac{1}{(2g-2)\sqrt{g+\delta}}$$

for $r \to \infty$. On the other hand a version of Dyson's Lemma [V3] yields asymptotically the upper bond $\sqrt{\frac{2\delta}{g+\delta}}$ for this index. If one chooses $\delta < \eta^2/8(g-1)^2$ then one obtains a contradiction. Therefore r is bounded and as a consequence h_2/h_1 is bounded and so is h_2.

In section 6 we shall show how to modify the proofs in order to establish the finiteness of rational points also in the case that B is the arithmetic curve Spec $\mathcal{O}_K$.

4 Arithmetic Riemann-Roch Theorem

In was natural to ask the question whether the theory of arithmetic surfaces initiated by Arakelov and Faltings could be generalized to higher dimensional varieties and to vector bundles instead of line bundles. In particular one of the basic questions was to prove an arithmetic version of the Grothendieck-Riemann-Roch theorem.

The first progress in this direction was the foundation of an arithmetic intersection theory for general varieties. This was done by Gillet and Soulé in their paper [GS1].

Then they introduced hermitian K-theory in [GS2]. Together with Bismut they started the search for a Riemann-Roch theorem for arithmetic schemes. It turned out that the obvious generalization of the classical Riemann-Roch theorem is false and a Riemann-Roch theorem for arithmetic varieties necessarily has to involve a certain secondary class $R(x)$. The first progress in this direction was made by Bismut and Lebeau [BL] who proved a Riemann-Roch result for closed immersion and then they derived the Riemann-Roch theorem for determinant bundles. For this they used stochastic integration. Recently Faltings [F1] was able to replace the stochastic methods by heat kernel methods and proved a general Riemann-Roch theorem conjectured by Gillet and Soulé in [GS3].

In the sequel we shall explain these results. For this we first have to introduce the Chow groups. They are defined for an arbitrary Noetherian regular scheme X over $\mathbb{Z}$ as follows. For integers $p \geq 0$ let $Z^p(X)$ be the free abelian group generated by all integral subvarieties of X of codimension p; this is the group of p-cycles. By $Z'^p(X)$ we denote the subgroup of $Z^p(X)$ generated by all p-cycles of the form $\operatorname{div}(f)$ for some meromorphic function f on some $p-1$-cycle in X. Then the Chow groups of X are

$$CH^p(X) = Z^p(X)/Z'^p(X)$$

and we put

$$\mathbf{CH}(X) = \bigoplus_{p\geq 0} CH^p(X)$$

and

$$\mathbf{CH}(X)_{\mathbb{Q}} = \mathbf{CH}(X) \otimes \mathbb{Q} \ .$$

The intersection of cycles defines a pairing

$$\mathbf{CH}(X)_{\mathbb{Q}} \times \mathbf{CH}(X)_{\mathbb{Q}} \rightarrow \mathbf{CH}(X)_{\mathbb{Q}}$$

which makes $\mathbf{CH}(X)_{\mathbb{Q}}$ into a commutative ring. Also if $f : X \rightarrow Y$ is a proper morphism one gets the direct image functor $f_* : CH^p(X) \rightarrow CH^{p-\delta}(Y)$ where δ is the fibre dimension. For a line bundle L on X and an integral subscheme Z of X let σ be a meromorphic section of L on Z and D be the Cartier divisor defined by σ. Then the class $[D]$ of D in $\mathbf{CH}(X)$ does not depend on the choice of σ. This class is the product $c_1(L) \cdot [Z]$ of the Chern class $c_1(L)$ of L and the class of Z.

The construction can be extended to regular schemes X of dimension $n+1$ which are flat over $\mathbb{Z}$. Then $X_{\mathbb{C}} = X \otimes \mathbb{C}$ is a smooth complex manifold of dimension n. For integers $p, q \geq 0$ we let $A^{p,q}(X_{\mathbb{C}})$ be the space of all (p,q)-forms with complex coefficients. Then we obtain as usual the differential operators

$$\partial : A^{p,q}(X_{\mathbb{C}}) \rightarrow A^{p+1,q}(X_{\mathbb{C}})$$
$$\bar{\partial} : A^{p,q}(X_{\mathbb{C}}) \rightarrow A^{p,q+1}(X_{\mathbb{C}})$$

and define

$$\tilde{A}^{p,q}(X_{\mathbb{C}}) = A^{p,q}(X_{\mathbb{C}})/(\operatorname{im}\partial + \operatorname{im}\bar{\partial}) \ .$$

We put these groups for $p = q$ together to obtain

$$\mathbf{A}(X_{\mathbb{C}}) = \bigoplus_{p \geq 0} A^{p,p}(X_{\mathbb{C}})$$

and

$$\tilde{\mathbf{A}}(X_{\mathbb{C}}) = \bigoplus_{p \geq 0} \tilde{A}^{p,p}(X_{\mathbb{C}}) \ .$$

The complex conjugation F on X induces an endomorphism F^* on $\mathbf{A}(X_{\mathbb{C}})$ and $\tilde{\mathbf{A}}(X_{\mathbb{C}})$ and we define

$$\mathbf{A}(X) = \{g = \Sigma g^p \in \mathbf{A}(X_{\mathbb{C}}), F^* g^p = (-1)^p g^p\}$$

and similarly $\tilde{\mathbf{A}}(X)$. We also need the extension of differential forms to currents. The space $D_{p,q}(X_{\mathbb{C}})$ is defined to be the space of linear functionals on $A^{p,q}(X_{\mathbb{C}})$ which are continuous in the sense of Schwartz. Then we put $D^{p,q}(X_{\mathbb{C}}) = D_{n-p,n-q}(X_{\mathbb{C}})$ and obtain an inclusion

$$A^{p,q}(X_{\mathbb{C}}) \hookrightarrow D^{p,q}(X_{\mathbb{C}})$$

given by $\omega \mapsto [\omega]$ where

$$[\omega](\alpha) := \int_X \omega \wedge \alpha, \quad \alpha \in A^{n-p,n-q}(X_{\mathbb{C}}) \ .$$

The operators ∂ and $\bar{\partial}$ extend to currents and the spaces $\mathbf{D}(X_{\mathbb{C}}), \tilde{\mathbf{D}}(X_{\mathbb{C}}), \mathbf{D}(X)$ are defined as above. Now we are ready to define the arithmetic Chow groups. Let Z be a p-cycle in $Z^p(X)$ and δ_Z be the current which gives integration on $Z_{\mathbb{C}}$. Then $Z^p_{\mathrm{Ar}}(X)$ is the group generated by pairs (Z, g) for $Z \in Z^p(X)$ and $g \in D^{n-p,n-p}(X)$ such that

$$h_Z = \delta_Z - \frac{\partial\bar{\partial}}{\pi i} g$$

is smooth modulo the subgroup generated by elements of type $(0, \partial\alpha + \bar{\partial}\beta)$ for some currents α, β. In $Z^p_{\mathrm{Ar}}(X)$ we have the subgroup $Z'^p_{\mathrm{Ar}}(X)$ generated by the elements

$$(\operatorname{div}(g), -\log|g|)$$

where g is a meromorphic function as above. Then the arithmetic Chow group is

$$CH^p_{\mathrm{Ar}}(X) = Z^p_{\mathrm{Ar}}(X)/Z'^p_{\mathrm{Ar}}(X)$$

and the Chow ring is

$$\mathbf{CH}_{\mathrm{Ar}}(X) = \bigoplus_{p \geq 0} CH^p_{\mathrm{Ar}}(X) \ .$$

The product on $\mathbf{CH}_{\mathrm{Ar}}(X)$ is induced by the star product

$$(Z,g)*(Z',g')=(Z\cdot Z',g\wedge\delta_{Z'}+h_Z\wedge g')$$

where the product $Z\cdot Z'$ is taken in $\mathbf{CH}_{\mathrm{Ar}}(X)$. On $CH^{n+1}_{\mathrm{Ar}}(X)$ we have a degree map given by

$$\begin{array}{ccl} CH^{n+1}_{\mathrm{Ar}}(X) & \xrightarrow{\deg} & \mathbb{R} \\ (Z,g) & \longmapsto & \deg(Z,g)=\log\sharp\Gamma(Z,O_Z)+\int\limits_{X_{\mathbb{C}}} g\ . \end{array}$$

Clearly this defines an intersection pairing

$$\begin{array}{ccl} CH^{p}_{\mathrm{Ar}}(X)\times CH^{n+1-p}_{\mathrm{Ar}}(X) \longrightarrow & & \mathbb{R} \\ \Big((Z,g),(Z',g')\Big) & \longmapsto & \deg((Z,g)*(Z',g'))\ . \end{array}$$

The first chern class of a hermitian line bundle (L,h) is defined by taking a section σ of L and putting

$$c_1(L,h)=(\mathrm{div}\ \sigma,-\log|\sigma|)\ .$$

This definition is independent of the choice of σ. Here we have put $|\sigma|^2=h(\sigma,\sigma)$.

Let $f:X\rightarrow Y$ be a proper morphism between regular schemes. Then f induces a homomorphism

$$f_{\mathbf{CH}_{\mathrm{Ar}}}:\mathbf{CH}_{\mathrm{Ar}}(X)\longrightarrow\mathbf{CH}_{\mathrm{Ar}}(Y)$$

between the arithmetic Chow groups. On cycles it is given by the direct image for cycles and on forms by fibrewise integration.

The next step is to define arithmetic K-theory. For this we first recall the classical K-theory for coherent (locally free) sheaves F on X. Let $\mathbf{G}(X)$ be the free abelian group generated by such sheaves F and $\mathbf{G}'(X)$ the subgroup generated by elements of the form $F-F'-F''$ where

$$\mathcal{F}:0\rightarrow F'\rightarrow F\rightarrow F''\rightarrow 0$$

is an exact sequence. Then

$$\mathbf{K}(X)=\mathbf{G}(X)/\mathbf{G}'(X)\ .$$

The elements of this group are denoted by $[F]$. Tensor product puts a commutative ring structure onto $\mathbf{K}(X)$. If $f:X\rightarrow Y$ is a proper morphism between regular schemes X and Y then again we obtain a direct image functor f_* from $\mathbf{K}(X)$ to $\mathbf{K}(Y)$ which maps F to $\sum\limits_{i\geq 0}(-1)^i[R^if_*F]$. Also a pull back morphism f^* from $\mathbf{K}(Y)$ to $\mathbf{K}(X)$ can be defined by

$$[F]\mapsto[f^*F]\ .$$

The functors f^* and f_* are related by $f_*[E \otimes f^*F] = f_*[E] \cdot [F]$. The Chow ring $\mathbf{CH}(X)$ and the K-group $\mathbf{K}(X)$ are related by the Chern classes of vector bundles. We have already introduced Chern classes of line bundles. The higher Chern classes $c_i(E)$ for vector bundles E are then defined by means of the splitting principle.

Further the Chern character $\mathbf{ch}(E)$ is given by

$$\mathbf{ch}(E) = \sum_i \exp\left(c_1(L_i)\right)$$

when after base change $E = \sum_i L_i$. It is a ring homomorphism

$$\mathbf{ch}(E) : \mathbf{K}(X) \to \mathbf{CH}(X)_{\mathbb{Q}}$$

and can be expressed as a polynomial in the $c_i(E)$ with rational coefficients. In order to state the classical Riemann-Roch theorem we also need the Todd class of E. Again this is a polynomial in the Chern classes $c_i(E)$ and given by

$$\mathbf{Td}(E) = \prod_i \frac{c_1(L_i)}{1 - e^{-c_1(L_i)}} \quad .$$

Then if $f : X \to Y$ is a projective and smooth morphism the Riemann-Roch theorem says that

$$f_*\left(\mathbf{ch}(E) \cdot \mathbf{Td}(T_{X/Y})\right) = \mathbf{ch}(f_*E)$$

in $\mathbf{CH}(Y)$. Here $T_{X/Y}$ is the relative tangent sheaf of X over Y.

The next step towards arithmetic K-groups is to imitate this construction for hermitian vector bundles. We start with a real manifold and consider the de Rham complex

$$(\Omega^*, d) : \Omega^0 \xrightarrow{d} \Omega^1 \xrightarrow{d} \Omega^2 \to \ldots$$

of complex smooth p-forms. Its cohomology is $H^*_{DR}(X, \mathbb{C})$. If X carries a Riemannian metric then so does Ω^*, the adjoint d^* of d can be defined and so we get the Laplacian

$$\Delta_d = dd^* + d^*d \quad .$$

If $\mathcal{H} = \mathrm{Ker}\Delta_d$ are the harmonic forms then by Hodge theory $\mathcal{H} \cong H^*_{DR}(X, \mathbb{C})$. When X is a complex manifold with hermitian metric then we get the double complex $(\Omega^{*,*}, \partial, \bar\partial)$ of (p,q)-forms $\Omega^{p,q}(X)$. The hermitian metric on X induces on $\Omega^{*,*}$ a hermitian metric and so ∂^* and $\bar\partial^*$ can be defined; this leads to the Laplacians

$$\Delta_\partial = \partial\partial^* + \partial^*\partial, \quad \Delta_{\bar\partial} = \bar\partial\bar\partial^* + \bar\partial^*\bar\partial \quad .$$

Let (E, h) be a hermitian vector bundle on X, $\nabla : E \to E \otimes \Omega^1$ a connection and $\nabla^2 \in \mathrm{End}(E) \otimes \Omega^2$ its curvature. Then the Chern polynomial $\mathrm{c}'_t(E, h)$ is defined as

$$\mathrm{c}'_t(E, h) = \det\left(t - \frac{i}{2\pi}\nabla^2\right) \quad .$$

The Chern character form is

$$\mathbf{ch}'(E,h) = \mathrm{tr}\ \exp\left(\frac{i}{2\pi}\nabla^2\right) \ .$$

In order to define the K-groups of hermitian vector bundles we have to study exact sequences of vector bundles which are also not necessarily exact sequences of hermitian vector bundles. This leads to the secondary Chern form which measures the defect of the sequence being exact sequence of hermitian vector bundles. If $\mathcal{E} : 0 \to E' \to E \to E'' \to 0$ is an exact sequence of vector bundles with hermitian metrics then the secondary Chern form $\widetilde{\mathbf{ch}}(\mathcal{E})$ is determined by

$$\frac{\partial\bar{\partial}}{\pi i}\widetilde{\mathbf{ch}}(\mathcal{E}) = \mathbf{ch}'(E,h) - \mathbf{ch}'(E',h') - \mathbf{ch}'(E'',h'') \ .$$

If $\mathcal{E}$ splits as an exact sequence of hermitian vector bundles then $\widetilde{\mathbf{ch}}(\mathcal{E}) = 0$. The arithmetic K-group $\mathbf{K}_{\mathrm{Ar}}(X)$ is defined for regular schemes X flat over $\mathbb{Z}$ in the following way: Let $\tilde{\mathbf{G}}(X)$ be the free group generated by hermitian vector bundles with hermitian metric invariant under F and put

$$\hat{\mathbf{G}}(X) = \tilde{\mathbf{G}}(X) \oplus \tilde{\mathbf{A}}(X) \ .$$

We form the subgroup $\hat{\mathbf{G}}'(X)$ of $\hat{\mathbf{G}}(X)$ generated by elements of the form

$$(E,h,0) - (E',h',0) - (E'',h'',0) - \widetilde{ch}(\mathcal{E})$$

with E, E', E'' as above. Then the arithmetic K-group $\mathbf{K}_{\mathrm{Ar}}(X)$ is given by

$$\mathbf{K}_{\mathrm{Ar}}(X) = \hat{\mathbf{G}}(X)/\hat{\mathbf{G}}'(X) \ .$$

It is now possible to define a homomorphism $\mathbf{ch}_{\mathrm{Ar}} : \mathbf{K}_{\mathrm{Ar}}(X)_{\mathbb{Q}} \to \mathbf{CH}_{\mathrm{Ar}}(X)_{\mathbb{Q}}$. For this let ϕ be a symmetric homogeneous polynomial in the variables $T_1, \ldots, T_n$ and A be any $\mathbb{R}$-algebra. Then ϕ extends uniquely to a polynomial map

$$\phi : M_n(A) \to A$$

from the matrix ring $M_n(A)$ into A. We associate to each hermitian vector bundle (E,h) the form $\phi(E,h) = \phi\left(\frac{i}{2\pi}\nabla^2\right)$ in $A(X)$. This construction extends to symmetric power series in $T_1, \ldots, T_n$. So in particular it is defined for Chern and Todd forms

$$\begin{aligned} \mathbf{ch}(T_1,\ldots,T_n) &= \sum_{i=1}^{n} \exp(T_i) \ , \\ \mathbf{Td}(T_1,\ldots,T_n) &= \prod_{i=1}^{n} \frac{T_i}{1-\exp(-T_i)} \ . \end{aligned}$$

The construction of the Bott-Chern secondary class $\widetilde{\mathbf{ch}}$ extends then to such power series ϕ and leads to the associated secondary class $\tilde{\phi}$ defined correspondingly.

Gillet and Soulé now extend this construction to attach to each element (E, h, η) in $\mathbf{K}_{\mathrm{Ar}}(X)$ and each ϕ as before an element $\phi_{\mathrm{Ar}}(E, h, \eta)$ in $\mathbf{CH}_{\mathrm{Ar}}(X)$. Their construction is again based on the first arithmetic Chern class $c_1(E, h, \eta)_{\mathrm{Ar}}$ defined as

$$c_1(E, h, \eta)_{\mathrm{Ar}} = c_1(E, h) + \varepsilon(\eta)$$

where $\varepsilon : \tilde{\mathbf{A}}(X) \to \mathbf{CH}_{\mathrm{Ar}}(X)$ is given by $g \mapsto \varepsilon(g) = (0, g)$. The final definition of ϕ_{Ar} is then obtained via a modification of the splitting principle using Grassmannians.

In particular for hermitian line bundles $(L_1, h_1), \ldots, (L_n, h_n)$ one has

$$\phi_{\mathrm{Ar}}\left((L_1, h_1) \oplus \ldots \oplus (L_n, h_n)\right) = \phi\left(c_1(L_1, h_1)_{\mathrm{Ar}}, \ldots, c_1(L_n, h_n)_{\mathrm{Ar}}\right)$$

and one gets the arithmetic Chern character

$$\mathbf{ch}_{\mathrm{Ar}} : \mathbf{K}_{\mathrm{Ar}}(X)_{\mathbb{Q}} \longrightarrow \mathbf{CH}_{\mathrm{Ar}}(X)_{\mathbb{Q}} \ .$$

Again $\mathbf{K}_{\mathrm{Ar}}(X)$ carries a ring sutructure induced by the tensor product. The crucial point now is do define the direct image functor $f_{\mathbf{K}_{\mathrm{Ar}}}$ on $\mathbf{K}_{\mathrm{Ar}}(X)$ with values in $\mathbf{K}_{\mathrm{Ar}}(Y)$ when $f : X \to Y$ is proper and generically smooth. One of the basic properties of $f_{\mathbf{K}_{\mathrm{Ar}}}$ is to respect the relation above which means that for an exact sequence $\mathcal{E} : 0 \to E' \to E \to E'' \to 0$ of acyclic bundles on X we should have in $\mathbf{K}_{\mathrm{Ar}}(Y)$ the relation

$$f_{\mathbf{K}_{\mathrm{Ar}}}(E, h, 0) = f_{\mathbf{K}_{\mathrm{Ar}}}(E', h', 0) + f_{\mathbf{K}_{\mathrm{Ar}}}(E'', h'', 0) + f_* \widetilde{\mathbf{ch}}(\mathcal{E}) \ .$$

In order to achieve this one has to modify the obvious definition of the direct image of an element (E, h, η) in $\hat{\mathbf{G}}(X)$ which would be

$$(E, h, \eta) \mapsto f_{\mathbf{K}_{\mathrm{Ar}}}(E, h, \eta) = (f_* E, f_* h, f_! \eta) \ ,$$

where $f_! \eta$ is given by

$$f_! \eta = f_* \left(\eta \, \mathbf{Td}(T_{X/Y}, h_{X/Y})\right) \ ,$$

by some element $\tau(E)$ in $\tilde{\mathbf{A}}(X)$. This element has to satisfy the functional equation

$$\tau(E) - \tau(E') - \tau(E'') - \widetilde{\mathbf{ch}}(f_* \mathcal{E}) = -f_! \left(\widetilde{\mathbf{ch}}(\mathcal{E})\right) \ .$$

The equation can be solved and one obtains the socalled higher analytic torsion. Thus the final definition for the direct image $f_{\mathbf{K}_{\mathrm{Ar}}}(E, h, \eta)$ is given by

$$f_{\mathbf{K}_{\mathrm{Ar}}}(E, h, \eta) = (f_* E, f_* h, \tau(E) + f_! \eta) \ .$$

Once the analytic torsion $\tau(E)$ has been defined for acyclic vector bundles the general result is obtained by resolving E by a complex of acyclic bundles.

Also we have to add a secondary class to the Todd-genus $\mathbf{Td}_{\mathrm{Ar}}(E,h)$. For every power series $R \in \mathbb{R}[[T]]$ one defines a characteristic class $R(E) \in \tilde{\mathbf{A}}(X)$ by the requirements that $R(f^*E) = f^*R(E)$, that $R(E)$ is additive on short exact sequences and that $R(L) = R(c_1(L))$ for line bundles. Define

$$\mathbf{Td}^R_{\mathrm{Ar}}(E,h) = \mathbf{Td}_{\mathrm{Ar}}(E,h)\,(1 - \varepsilon\,(R(E,h)))\ .$$

Then we can formulate the Riemann-Roch theorem.

4.1 Arithmetic Riemann-Roch Theorem ([F1]). *There exists a unique power series $R(T)$ such that*

$$\mathbf{ch}_{\mathrm{Ar}}\left(f_{\mathrm{K}_{\mathrm{Ar}}}(E,h,\eta)\right) = f_{\mathrm{CH}_{\mathrm{Ar}}}\left(\mathbf{ch}_{\mathrm{Ar}}(E,h,\eta)\cdot \mathbf{Td}^R_{\mathrm{Ar}}(T_{X/Y},h_{X/Y})\right)\ .$$

The proof of this theorem is extremely technical and involved. Deep methods from elliptic operator theory about heat kernels have to be used for spin bundles, Dirac operators and super connections.

5 Applications in Arithmetic

One of the main applications of the classical Riemann-Roch-Theorem is to establish the existence of sections of a vector bundle E on a projective variety X defined over the field of complex numbers. It allows to compute the dimension of $H^0(X,E)$ in many cases. In particular the growth of the dimension of $H^0(X, E\otimes L^t)$ can be determined when L is an ample line bundle and t a positive integer which tends to infinity. For an arithmetic version of Riemann-Roch one considers regular schemes X flat over spec $\mathbb{Z}$. Let E be a vector bundle and L an ample line bundle over X as before. Then one is interested in $H^0(X, E\otimes L^t)$ which is a finitely generated module M over $\mathbb{Z}$. It induces a lattice M' in $M\otimes_{\mathbb{Z}}\mathbb{R}$ and sections of M' are lattice points in this lattice. If $M\otimes_{\mathbb{Z}}\mathbb{R}$ carries an euclidian structure one can apply the geometry of numbers in order to determine small lattice points, a basis of M' and so on in terms of the given lattice datas.

Let K be a number field of degree d over the rationals and $\mathcal{O}_K$ its ring of integers. We denote by K_v the completion of K and by $\mathcal{O}_v$ the completion of $\mathcal{O}_K$ at the place v of K. The adele ring of K is denoted by $K_{\mathbb{A}}$ and we have a canonical injection φ from K into $K_{\mathbb{A}}$. On $K_{\mathbb{A}}$ we can define a standard Haar measure $\beta = \prod \beta_v$ by putting

$$\begin{array}{rclcl} \beta_v(\mathcal{O}_K) & = & 1 & , & v \nmid \infty \\ d\beta_v & = & dx & , & v \text{ real} \\ d\beta_v & = & |dz \wedge d\bar{z}| \;=\; -2i\, du \wedge dv, & & v \text{ complex,} \end{array}$$

where $z = u + iv$. Then it is known [We] that the quotient $K_{\mathbb{A}}/K$ has volume $|D_K|^{1/2}$ where D_K is the discriminant of K.

Let M be a hermitian $\mathcal{O}_K$-module of rank n without torsion. This is an $\mathcal{O}_K$-module M with hermitian metrics h_σ on $M \otimes_\sigma \mathbb{C}$ for each $\sigma \in \Sigma = \mathrm{Hom}(K, \mathbb{C})$. Complex conjugation operates on Σ and induces a $\mathbb{C}$-antilinear involution F_∞ on $\mathbb{C}^\Sigma$ which fixes the image of M in $M^\Sigma = \coprod M \otimes_\sigma \mathbb{C}$; here the product is taken over all $\sigma \in \Sigma$. Let E be the K-vector space $M \otimes_{\mathcal{O}_K} K$ and E_∞ the subspace of M^Σ fixed by F_∞. Then clearly $E_\infty = \prod E_v$ where E_v is the completion of E at v and the product is over all $v \mid \infty$ which is the same as $E \otimes_K K_v$ and E_∞ is also isomorphic to $E \otimes_{\mathbb{Q}} \mathbb{R}$. The hermitian metrics h_σ induce a hermitian form h on M^Σ which restricts to an euclidian scalar product $\varepsilon = \prod \varepsilon_v$ on E_∞ where again the product is over the places as above. We fix now an isomorphism $\Theta : K^n \xrightarrow{\sim} E$. Clearly Θ extends to an isomorphism

$$\Theta : K^n_{\mathbb{A}} \longrightarrow E_{\mathbb{A}}$$

where $E_{\mathbb{A}} = \prod E_v \cong E \otimes_K K_{\mathbb{A}}$ and the product is taken over all v. Let $\Lambda = \Theta^{-1}M$. This is an $\mathcal{O}_K$-lattice in K^n and induces a lattice Λ_v in K^n_v for each $v \nmid \infty$. In $K^n_\infty = K^n \otimes_{\mathbb{Q}} \mathbb{R} = \prod K^n_v$ we define the convex body $\mathcal{B}$ as the inverse image under Θ of the convex set in E_∞ given by $\varepsilon < 1$. In this way we obtain the set

$$\mathcal{S} = \mathcal{B} \times \prod_{v \nmid \infty} \Lambda_v$$

whose closure is compact in $K^n_{\mathbb{A}}$. Then in the notations of [Bo-Va] the successive minima for $\mathcal{S}$ satisfy

$$(\lambda_1 \ldots \lambda_n)^d \beta(\mathcal{S}) \leq 2^{nd} \beta(K^n_{\mathbb{A}}/K^n) \quad .$$

Note that our normalization of β is different to that in [Bo-Va] so that the right hand side becomes $|D_K|^{n/2}$ instead of 1. Since Λ is a lattice in K^n the quantity $\chi(\Lambda, \mathcal{O}^n_K)$ can be defined as in [S4], p. 47. It is a fractional ideal in $\mathcal{O}_K$ so that

$$|\chi(\Lambda, \mathcal{O}^n_K)| = \sharp \mathcal{O}_K / \chi(\Lambda, \mathcal{O}^n_K)$$

is defined. Then as a consequence of our normalizations

$$\prod_{v \nmid \infty} \beta_v(\Lambda) = |\chi(\Lambda, \mathcal{O}^n_K)| \quad .$$

Furthermore the volume of $\mathcal{B}$ is given by

$$\prod_{v \mid \infty} \beta_v(\mathcal{B}) = \prod_{\sigma} \det h_\sigma(e_i, e_j)^{-1/2} \cdot \prod_{v \mid \infty} \beta_v(\rho_v)$$

where e_i is the image of the i-th standard basis vector in K^n under Θ and ρ_v is the euclidean unit ball in K_v^n. We put $\mu_\infty = \prod \beta_v(\rho_v)$. Then the above formula on successive minima becomes

$$(\lambda_1 \dots \lambda_n)^d \leq 2^{nd} \mu_\infty^{-1} |D_K|^{n/2} \exp(-\chi(M; h))$$

where we put

$$\chi(M; h) = -\frac{1}{2} \log \prod \det h_\sigma(e_i, e_j) + \log |\chi(\Lambda, \mathcal{O}_K^n)| \quad .$$

We observe that the right hand side does not depend on the choice of the isomorphism Θ but only on M and h. We call $\chi(M; h)$ the Euler characteristic of (M, h).

If M is a hermitian $\mathcal{O}_K$-module with torsion and M' is the torsionfree part of M then we write $M = M' \oplus M_{tor}$ where M_{tor} is the torsion and define

$$\chi(M; h) = \chi(M'; h) + \log \sharp M_{tor} \quad .$$

Thus the determination of the Euler Characteristic is an important matter for finding small elements in M.

Hermitian modules naturally arise in arithmetic. Let $\pi: X \rightarrow \operatorname{Spec} \mathcal{O}_K$ be a projective regular flat scheme of relative dimension n, (E, h_E) a hermitian vector bundle. We fix a Kähler form with associated Kähler metric ω_0 on $X^\Sigma = \coprod X_\sigma = X \otimes_{\mathcal{O}_K} K^\Sigma$ where $X_\sigma = X \otimes_\sigma \mathbb{C}$ and the disjoint union is over all $\sigma \in \Sigma$. We put $X_\infty = X^\Sigma(\mathbb{C})$. For a section σ of E we can define then the L^2-norm

$$||\sigma||_{L^2}^2 = \int_{X_{\mathbb{C}}} h_E(\sigma, \sigma) \frac{\omega_0^n}{n!} \quad .$$

This norm extends naturally to the Dolbeault complex so that for sections ω of $E \otimes \Omega^{0,q}$ the norm $||\omega||_{L^2}$ is defined. As a consequence one gets the $\bar{\partial}$-Laplacian $\Delta_{\bar{\partial}}$. On noting that the de Rham cohomology is isomorphic to the space of harmonic forms by Hodge theory one obtains a hermitian form on the direct image $\pi_* E$ which we denote by $\pi_* h_E$. Furthermore we have seen in section 4 that the direct image $\pi_{K_{Ar}}(E, h_E, 0)$ is given by the triple $(\pi_* E, \pi_* h_E, \tau(E))$ where $\tau(E)$ is the analytic torsion. Since the complex dimension of Spec $\mathcal{O}_K$ is zero, only the component $\tau_0(E)$ of degree 0 of $\tau(E)$ matters and this component can be calculated. To do this let $\lambda_{1,q} \leq \lambda_{2,q} \leq \dots$ be the positive eigenvalues of $\Delta_{\bar{\partial}}$ and define the Zeta-function ζ_q as

$$\zeta_q(s) = \sum_{n \geq 1} \lambda_{n,q}^{-s} \quad .$$

It converges absolutely for Re $s >> 1$, it can be continued meromorphically to the whole complex plane and is holomorphic at $s = 0$. Then the analytic torsion is given by

$$\tau_0(E) = \sum_{q>0} (-1)^{q+1} q \zeta_q'(0) \ .$$

The hermitian $\mathcal{O}_K$-modules to which we apply the arithmetic Riemann-Roch are then the modules

$$(\pi_* E, \pi_* h_E)$$

and its determinant

$$\det(\pi_* E, \pi_* h_E)$$

which is defined to be the pair

$$(\lambda(E), h_{L^2})$$

with

$$\lambda(E) = \bigotimes_q \Lambda^{\max} H^q(X, E)^{(-1)^q} \ .$$

Further h_{L^2} is the metric on $\lambda(E)$ induced by $\pi_* h_E$. Clearly

$$\chi(\pi_* E; \pi_* h_E) = \chi\left(\lambda(E); h_{L^2}\right)$$

and, if we define the Quillen metric h_Q on $\lambda(E)$ by twisting with the analytic torsion

$$h_Q = h_{L^2} \exp\left(-\tau_0(E)\right) \ ,$$

one gets

$$\chi\left(\lambda(E); h_Q\right) = \chi\left(\lambda(E); h_{L^2}\right) - \tau_0(E) \ .$$

Next we shall describe the arithmetic Chow group in our arithmetic situation. The morphism π induces a homomorphism $\pi_{\mathbf{CH}_{\mathrm{Ar}}}$ between the Chow groups $\mathbf{CH}_{\mathrm{Ar}}(X)$ and $\mathbf{CH}_{\mathrm{Ar}}(\mathrm{Spec}\ \mathcal{O}_K)$. The non-trivial parts are

$$f_{\mathbf{CH}_{\mathrm{Ar}}} : CH^n_{\mathrm{Ar}}(X) \longrightarrow CH^0_{\mathrm{Ar}}(\mathrm{Spec}\ \mathcal{O}_K)$$

and

$$f_{\mathbf{CH}_{\mathrm{Ar}}} : CH^{n+1}_{\mathrm{Ar}}(X) \longrightarrow CH^1_{\mathrm{Ar}}(\mathrm{Spec}\ \mathcal{O}_K) \ .$$

The group $CH^0_{\mathrm{Ar}}(\mathrm{Spec}\ \mathcal{O}_K)$ is the free abelian group generated by the connected components of Spec $\mathcal{O}_K$. The group $CH^1_{\mathrm{Ar}}(\mathrm{Spec}\ \mathcal{O}_K)$ can be described by the exact sequence

$$1 \to \mu(\mathcal{O}_K) \to \mathcal{O}_K^* \xrightarrow{\rho} \mathbb{R}^{r_1+r_2} \xrightarrow{a} CH^1_{\mathrm{Ar}}(\mathrm{Spec}\ \mathcal{O}_K) \xrightarrow{\zeta} Cl(\mathcal{O}_K) \to 1$$

where $\mu(\mathcal{O}_K)$ are the roots of unity, $\mathcal{O}_K^*$ the units of $\mathcal{O}_K$ and r_1 (resp. r_2) the real (resp. complex) embeddings of $\mathcal{O}_K$. Further $Cl(\mathcal{O}_K)$ is the class group of $\mathcal{O}_K$. The maps are given by

$$\rho : \alpha \mapsto (\ldots, \log|\sigma\alpha|, \ldots) \ ,$$

$a : g \mapsto (0, g)$ and $\zeta : (Z, g) \mapsto Z$. There is a homomorphism

$$\deg : CH^1_{\mathrm{Ar}}(\mathrm{Spec}\, \mathcal{O}_K) \to \mathbb{R}$$

which was defined in section 4.

Both, the arithmetic K-group $\mathbf{K}_{\mathrm{Ar}}(\mathrm{Spec}\, \mathcal{O}_K)$ and the arithmetic Chow group $\mathbf{CH}_{\mathrm{Ar}}(\mathrm{Spec}\, \mathcal{O}_K)$ are related by the arithmetic Chern character

$$\mathbf{ch}_{\mathrm{Ar}} : \mathbf{K}_{\mathrm{Ar}}(\mathrm{Spec}\, \mathcal{O}_K) \to \mathbf{CH}_{\mathrm{Ar}}(\mathrm{Spec}\, \mathcal{O}_K)_{\mathbb{Q}} \quad .$$

In particular for arithmetic line sheaves (L, h, η) we have

$$\deg\left(c_1(L, h, \eta)_{\mathrm{Ar}}\right) = \chi(L; h) + \eta = \chi(L; e^{-\eta} h) \quad ,$$

as can be easily verified. Furthermore for an arbitrary hermitian module (E, h, η) we have

$$c_1(E, h, \eta)_{\mathrm{Ar}} = c_1(\det E, \det h, \eta)_{\mathrm{Ar}}$$

where $\det E$ is the maximal exterior power of E and $\det h$ its corresponding hermitian form. Therefore we conclude that

$$\deg\left(c_1(E, h, \eta)_{\mathrm{Ar}}\right) = \chi(\det E; e^{-\eta} \det h) \quad .$$

If this is applied to the direct image as above one obtains

$$\chi\left(\lambda(E); e^{-\eta} h_Q\right) = \deg\left(\mathbf{ch}_{\mathrm{Ar}}\left(\pi_{\mathrm{K}_{\mathrm{Ar}}}(E, h, \eta)\right)\right) \quad .$$

The term on the right hand side is by the arithmetic Riemann-Roch theorem equal to

$$\deg \pi_{\mathrm{CH}_{\mathrm{Ar}}}\left(\mathbf{ch}_{\mathrm{Ar}}(E, h, \eta) \cdot \mathbf{Td}^R_{\mathrm{Ar}}(T_X, \omega_0)\right) \quad .$$

Thus we derive the following version of the arithmetic Riemann-Roch theorem.

5.1 Theorem. *Let $\pi : X \to \mathrm{Spec}\, \mathcal{O}_K$ be a regular and projective scheme flat over $\mathrm{Spec}\, \mathcal{O}_K$. Then we have for $(E, h, \eta) \in \mathbf{K}_{\mathrm{Ar}}(X)$*

$$\chi\left(\lambda(E); e^{-\eta} h_Q\right) = \deg \pi_{\mathrm{CH}_{\mathrm{Ar}}}\left(\mathbf{ch}_{\mathrm{Ar}}(E, h, \eta) \cdot \mathbf{Td}^R_{\mathrm{Ar}}(T_X, \omega_0)\right) \quad .$$

6 Small Sections

We indicate in this section how Theorem 5.1 can be used to find small sections of vector bundles. For this let $X_{\mathrm{Ar}} = (X, \omega_0)$ be an Arakelov variety over the ring of integers of a number field K. This is a pair consisting of a regular scheme X over $\mathcal{O}_K$ with projective generic fibre and a Kähler form ω_0 on $X_{\mathbb{C}}$ which satisfies $F_\infty^*\omega_0 = -\omega_0$. Let $\mathcal{H}(X)$ denote denote the space of harmonic forms in $\mathbf{A}(X)$ and let $\omega : \mathbf{CH}_{\mathrm{Ar}}(X) \to \mathbf{A}(X)$ be the projection which maps (Z,g) onto $h_Z = \delta_Z - \frac{\partial\bar\partial}{\pi i}g$. Then we define $\mathbf{CH}_{\mathrm{Ar}}(X_{\mathrm{Ar}})$ to be the inverse image of $\mathcal{H}(X)$ under ω. The intersection product on $\mathbf{CH}_{\mathrm{Ar}}(X)$ restricts to an intersection product on $\mathbf{CH}_{\mathrm{Ar}}(X_{\mathrm{Ar}})$. If $n_1,\ldots,n_r$ are non-negative integers with $n_1+\ldots+n_r = \dim X = n+1$ we define a multilinear intersection pairing

$$CH^{n_1}_{\mathrm{Ar}}(X_{\mathrm{Ar}}) \otimes \ldots \otimes CH^{n_r}_{\mathrm{Ar}}(X_{\mathrm{Ar}}) \to CH^{n+1}_{\mathrm{Ar}}(X_{\mathrm{Ar}}) \xrightarrow{\deg} \mathbb{R} \ .$$

For $n=1$ this is the same as the Arakelov intersection pairing defined in Arakelov theory for arithmetic surfaces. Here $n_1 = n_2 = 1$.

Let (E,h_E) be a hermitian vector bundle and (L,h_L) a positive holomorphic ample hermitian line bundle on X. There exists a unique holomorphic hermitian connection ∇ on (L,h_L); its curvature $R=\nabla^2$ is a section of the bundle $\Omega^{1,1}(X)\otimes$ End L. Then by definition (L,h_L) is positive if for any non-zero holomorphic tangent vector u on X we have $R(u,\bar u) > 0$. We apply as an example Theorem 5.1 in this situation to find a small section of $E\otimes L^t$ for sufficiently large t. Let h be the metric induced by h_E and h_L on $E\otimes L^t$. Since L is ample it follows that $H^q(X,E\otimes L^t) = 0$ for $q>0$ and sufficiently large t. Therefore with respect to the Quillen metric h_Q induced by h on $\lambda(E\otimes L^t)$ we have

$$\chi\left(\lambda(E\otimes L^t); h_Q\right) = \chi\left(\det H^0(X,E\otimes L^t); h_Q\right) \ .$$

By a result of Bismut and Vasserot [BV] (see also [F1]) the analytic torsion $\tau_0(E\otimes L^t)$ has an asymptotic expansion

$$\tau_0(E\otimes L^t) = \frac{\text{rank } E}{2n!} t^n \log t + O(t^n) \ .$$

Therefore

$$\chi\left(\det H^0(X,E\otimes L^t); h_Q\right) = \chi\left(\det H^0(X,E\otimes L^t), h_{L^2}\right) + O(t^n \log t) \ .$$

On the other hand the right hand side of the arithmetic Riemann-Roch formula becomes

$$\text{rank } E \cdot c_1(L,h_L)^{n+1}_{\mathrm{Ar}} \frac{t^{n+1}}{(n+1)!} + O(t^n) \ .$$

Here $c_1(L,h_L)^{n+1}_{\mathrm{Ar}}$ is the above intersection product. When we put everything together we obtain the following result (see [GS4]):

6.1 Theorem. *Under the above assumptions*

$$\chi\left(H^0(X,E\otimes L^t); h_{L^2}\right) = \text{rank } E \cdot c_1(L,h_L)^{n+1}_{\mathrm{Ar}} \frac{t^{n+1}}{(n+1)!} + O(t^n \log t) \ .$$

This result can be used in order to obtain an upper bound for the first successive minimum λ_1. We estimate the right hand side of the Minkowski formula indicated above. For this we need to calculate μ_∞ and since the size of μ_∞ depends on the rank of $H^0(X, E \otimes L^t)$ we get by the classical Riemann-Roch theorem the expansion

$$\text{rank } E \cdot c_1(L, h_L)^n \frac{t^n}{n!} + O(t^{n-1})$$

for this dimension. Thus by Stirling's formula we find that (recall that $d = [K : \mathbb{Q}]$)

$$\log \mu_\infty = d\, n\, t^n \log t + O(t^n) \ .$$

This leads to

$$\log \lambda_1 \leq -\frac{1}{d} \frac{c_1(L, h_L)^{n+1}_{\text{Ar}}}{c_1(L, h_L)^n} \frac{t}{n+1} + \gamma \cdot \log t$$

with some positive constant γ. It follows that there exists a section s of $E \otimes L^t$ such that for all $\sigma \in \Sigma$

$$||s||^2_{L^2, \sigma} = \int\limits_{X_\sigma} h_\sigma(s, s) \frac{\omega^n_{0,\sigma}}{n!} < \exp(-\kappa t)$$

for some constant $\kappa > 0$. Here h_σ and $\omega_{0,\sigma}$ are the restictions of h and ω_0 to X_σ.

6.2 Theorem. *Under the above assumptions for t sufficiently large there exists a constant $\kappa > 0$ and a section s of $E \otimes L^t$ such that for all $\sigma \in \Sigma$*

$$||s||^2_{L^2, \sigma} < \exp(-\kappa t) \ .$$

In the applications in diophantine geometry the L^2-norm is not very useful. Here one needs estimates for the supremum norm. Since the underlying manifolds are compact both norms are comparable and so the L^2-norm can be replaced by the supremum norm. This was pointed out by Gromov.

7 Vojta's Proof in the Number Field Case

In this section we give an exposition of Vojta's proof of Mordell's conjecture in the number field case. We keep the notations from section 3. In particular $B = \text{Spec } \mathcal{O}_K$ and we have constructed the divisor

$$\tilde{V}_r = \Delta' + a_1 F_1 + a_2 F_2 + bF$$

from which we know that for d sufficiently large the divisor $d\tilde{V}_r$ is linearly equivalent to an effective arithmetic divisor V_{Ar}. One of the main steps is now to construct a small section of $\mathcal{O}(d\tilde{V}_r)$. This is done in a similar way as in section 6 using the arithmetic Riemann-Roch. In the present situation however one does not know that $\mathcal{O}(d\tilde{V}_r)$ is ample over B and so its cohomology might have torsion. Therefore the hermitian modules $R^q\pi_*\mathcal{O}(d\tilde{V}_r)$ are not necessarily zero for $q > 0$. Since these sheaves are torsion sheaves for $q > 0$ and d sufficiently large clearly their Euler characteristics are non-negative. It follows from this and the fact that $\dim W = 2$ that one needs only to take care of $R^2\pi_*\mathcal{O}(d\tilde{V}_r)$. For this sheaf one shows that

$$\chi\left(R^2\pi_*\mathcal{O}(d\tilde{V}_r);||\ \ ||_{L^2}\right) = O(d^2)$$

where $||\ \ ||_{L^2}$ is the norm induced by the norms on the components of $\tilde{V}_r$. Therefore we again derive from Theorem 6.1 that

$$\chi\left(H^0\left(W;\mathcal{O}(d\tilde{V}_r)\right);||\ \ ||_{L^2}\right) = c_1\left(\mathcal{O}(d\tilde{V}_r),||\ \ ||\right)^3_{\text{Ar}}\frac{d^3}{6} + \mathcal{O}(d^2\log d)$$

and, since one can show that the leading term of this expansion is positive, that there exists a section s such that for some $\kappa > 0$

$$||s||^2_{L^2,\sigma} < \exp(-\kappa d)$$

for all $\sigma \in \Sigma$ and d sufficiently large. We next show that this section has a large index with respect to da_1, da_2 at a properly chosen point (P_1, P_2) where $P_i \in C, i = 1, 2$. We let E_i be the section of $X \to B$ corresponding to P_i, take E as in section 3 and let h_i be the (logarithmic) heights of E_i for $i = 1, 2$. We choose P_1, P_2 in such a way that $h_2 < rh_1$ and $\sqrt{h_2} \geq \sqrt{r}\sqrt{h_1}(1-\mu)$ for some positive μ. Then one shows again that the arithmetic intersection number of $d\tilde{V}_r$ and E satisfies

$$d\tilde{V}_r \cdot E \leq -\gamma \cdot d\sqrt{r}h_1$$

for some $\gamma > 0$. Let s be the section of $d\tilde{V}_r$ constructed above. We want to show that s vanishes along E (eventually to high order weighted by da_1, da_2 as above). Since s is an integral section the intersection number above is bounded from below by the terms coming from the places at infinity provided s does not vanish along E. Thus

$$d\tilde{V}_r \cdot E \geq -\sum_\sigma \log||s||_{\text{sup},\sigma}$$

where $||s||_{\text{sup}}$ is the supremum norm. Unfortunately the section is bounded only in terms of the L^2-norm. However since the manifold is compact one can compare the two norms. Indeed one shows that

$$\sum_\sigma \log||s||_{\text{sup},\sigma} \leq \sum_\sigma(\log d + \log||s||_{L^2,\sigma}) + c$$

for some constant c. The right hand side is bounded from above by $-\kappa' d$ for some positive constant κ' and d sufficiently large by the construction of s. Therefore we get together

$$\kappa' d \leq d\tilde{V}_r \cdot E \leq -\gamma d\sqrt{r} h_1$$

for d large. This is a contradiction and therefore s vanishes along E as claimed. If one analyzes the final inequality leading to the contradiction one sees that there is plenty of further space. In particular the contradiction does not depend on any special conditions on h_1. This observation infact is used in the next step to get more vanishing and to show that the index of the section y at (P_1, P_2) is larger than allowed to be according to Dyson's lemma. This leads then to ineffective bounds for the heights of the rational points on C and gives the finiteness of rational points.

8 Lang's Conjectures

Let $K \supseteq \mathbb{Q}$ be a field of finite type and A an abelian variety over K. Consider a subvariety $X \subseteq A$ over K. Then it is natural to determine the set $X(K)$ of K-rational points. Clearly if $B \subseteq A$ is any abelian subvariety of A such that $\xi + B \subseteq X$ for some $\xi \in A(\mathbb{C})$ then $(\xi + B)(K) \subseteq X(K)$. Hence if the left hand side is non-empty then $\xi \in X(K)$ and so $X(K)$ contains a whole coset of $B(K)$. In 1960 Lang [L2] conjectured that this is the general structure. More precisely he made the following two conjectures

Conjecture I: *$X(K)$ is contained in a finite number of translates of abelian subvarieties lying in X.*

Let $\Gamma_0 \subseteq A(\mathbb{C})$ be a finitely generated subgroup and $\Gamma \supseteq \Gamma_0$ its division group, i.e.,

$$\Gamma = \{\gamma \in A(\mathbb{C}); \mathbb{Z} \cdot \gamma \cap \Gamma_0 \neq 0\} \ .$$

Conjecture II: *$X(\mathbb{C}) \cap \Gamma$ is contained in a finite number of translates of abelian subvarietes lying in X.*

Let us consider first the special case that X does not contain a translate of an abelian subvariety of A. Then both conjectures are related. This was shown by Raynaud ([Ra]).

Theorem (Raynaud). *Conjecture II is implied by Conjecture I.*

Conjecture I was proved by Faltings [F2] in the case when K is a number field. So we have the following theorem.

Theorem (Faltings, 1989). *If X does not contain a translate of an abelian subvariety then $X(K)$ is finite.*

We now drop the hypothesis that X does not contain a translate of an abelian subvariety. Also this case was solved recently by Faltings [F3].

Theorem (Faltings, 1990). *Let K be a field of finite type over $\mathbb{Q}$ and X a subvariety of A. Then $X(K)$ is contained in a finite number of translates of abelian subvarieties lying in X.*

The proof of this theorem bases heavily on a result of Kawamata ([K]). Let $S(X)$ be the connected component of the neutral element of the stabilizer of X.

Kawamata's Structure Theorem: *Suppose that $S(X) = 0$. Then there exist an integer $\ell \geq 0$ and subvarieties $Z_1, \ldots, Z_\ell$ such that $S(Z_j) \neq 0$ for $j = 1, \ldots, \ell$ with the following property. Any translate of an abelian subvariety $B \neq 0$ of A which is a subvariety of X is contained in $Z(X) = Z_1 \cup \ldots \cup Z_\ell$.*

We note that Faltings' general theorem follows easily once one can show that the set $(X \setminus Z(X))(K)$ is finite. If this is the case then one has only to show that $Z(X)(K)$ is contained in a finite number of translates of abelian subvarieties as specified above. This follows by induction on the dimension. Namely if Z is a component of $Z(X)$ then $S(Z) \neq 0$. If we replace Z by $Z' = Z/S(Z)$ then $Z'(K)$ is contained in a finite number of translates and therefore $Z(X)(K)$ has the same property.

The finiteness of $(X \setminus Z(X))(K)$ follows up to some technical complications in the same way as the finiteness of $X(K)$ in [F2].

The techniques developed for the proof of Conjecture I lead also to a proof of another Conjecture of S. Lang namely that the complement of an ample divisor on an abelian variety over a number field contains only finitely many integral points. More generally we consider an abelian variety A over a number field K and an ample line bundle on A. Accordingly one can define the logarithmic height $h_L(x)$

of a K-rational point x on A. Let E be a K-subvariety of A and v a place of K. Then one defines the logarithmic v-adic distance $d_v(x, E)$ from x to E. In [F2] the following result is shown.

Theorem (Faltings, 1989). *For given $\varepsilon > 0$ there are only finitely many $x \in (A \setminus E)(K)$ such that*

$$d_v(x, E) < -\varepsilon h_L(x) \ .$$

One can ask for related results when A is replaced by a semiabelian variety. In this direction just recently Vojta presented a paper [V4] where he proves the analogue of Faltings' result on rational points on subvarieties of abelian varieties for integral points on subvarieties of semiabelian varieties.

9 Proof of Faltings' Theorem

We discuss now the proof of Faltings' theorem on rational points on subvarieties of abelian varieties but we shall restrict ourselves to the case when X does not contain any translate of an abelian subvariety. Furthermore we shall concentrate ourselves to present the four main innovations which are the key for obtaining the result. The first device is Faltings' line bundle which is the translation of Vojta's divisor to the present situation. If $j : C \times C \to \mathcal{J}(C) \times \mathcal{J}(C)$ is the embedding of $C \times C$ into the product $\mathcal{J}(C) \times \mathcal{J}(C)$ of the Jacobians of the curve C then Vojta's divisor is just the pull back of the Faltings divisor under the map j. The second is the introduction of a new height in projective and multiprojective space which is obtained by arithmetic intersection theory. In this situation the theory is much easier since there are not many line bundles on the projective space. The third new ingredient is the very important product theorem and finally the arithmetic Riemann Roch is replaced by a simple technique which uses directly the geometry of numbers. In this way all the difficult and highly elaborated techniques which were developed in the last years in this field get eliminated.

Let A be an abelian variety over K and $\mu : A \times A \to A$ the addition, $p_i : A \times A \to A, i = 1, 2$, the projection, L a very ample symmetric line bundle over A.

For integers s, t we consider the morphism $sp_1 + tp_2$. By pull-back with this morphism we obtain on $A \times A$ the line bundle $(sp_1 + tp_2)^* L$. More generally let s, t be rational numbers and n a positive integer such that ns, nt are integers. Then we let $(sp_1 + tp_2)^* L$ be the line bundle $n^{-2}(nsp_1 + ntp_2)^* L$. Note that this is not necessarily defined over K but as we shall work with a sufficiently high divisible multiple of it this does not matter since in this way the denominator in the exponent is cleared automatically. We let $\mathcal{P} = \mu^* L - p_1^* L - p_2^* L$ be the Poincaré bundle. Then

$$(sp_1 + tp_2)^* L = s^2 p_1^* L + t^2 p_2^* L + st\mathcal{P} \ .$$

For positive rational numbers $\varepsilon, s_1, \ldots, s_m$ we define the line bundles

$$\mathcal{G}_\pm = \pm\varepsilon(s_1^2 p_1^* L + \ldots + s_m^2 p_m^* L) + \sum_{i=1}^{m-1} (s_i p_i \pm s_{i+1} p_{i+1})^* L$$

and with the obvious notation

$$\mathcal{F}_\pm = (2 \pm \varepsilon) \sum_{i=1}^{m} s_i^2 p_i^* L \pm \sum_{i=1}^{m-1} s_i s_{i+1} \mathcal{P}_{i,i+1} \ .$$

Then

$$\mathcal{F}_\pm = \mathcal{G}_\pm + s_1^2 p_1^* L + s_m^2 p_m^* L \ .$$

There exists a positive number s such that for $s_i/s_{i+1} \geq s$, $1 \leq i \leq m-1$, the bundles $\mathcal{G}_\pm$ restrict to ample bundles on X^m provided m is large and ε sufficiently small. Since furthermore the bundle $s_1^2 p_1^* L + s_m^2 p_m^* L$ is generated by global sections also $\mathcal{F}_\pm$ are ample on X^m. They have the property that

$$\mathcal{F}_+ = 4\Sigma s_i^2 p_i^* L - \mathcal{F}_- = \mathrm{Hom}(\mathcal{F}_-, 4\Sigma s_i^2 p_i^* L) \ .$$

We choose now an integer d sufficiently large and consider the d-fold multiples of these line bundles. Since $\mathcal{F}_+$ is ample $d\mathcal{F}_+$ is very ample and so there exist injections $\rho_\lambda, \lambda = 1, \ldots, \ell$, of $d\mathcal{F}_-$ into $4d\Sigma s_i^2 p_i^* L$ without any common zero on X. We put $d_i = 4s_i^2 d$. Then the ρ_λ define an injection

$$\rho : \mathcal{F}_-^{\otimes d} \rightarrow \bigoplus_{\lambda=1}^{\ell} \bigotimes_{i=1}^{m} p_i^* L^{\otimes d_i}$$

given by $f \mapsto (\rho_1(f), \ldots, \rho_\ell(f))$. In such a situation, namely that $\mathcal{L}$ is a line bundle, $\mathcal{M}$ a locally free sheaf and $\rho : \mathcal{L} \to \mathcal{M}$ is an injection without a zero on X^m, one can define in a natural way two homomorphisms

$$\mathcal{M} \overset{\varepsilon_1}{\underset{\varepsilon_2}{\rightrightarrows}} \mathcal{M} \otimes \mathcal{M} \otimes \mathcal{L}^{-1} = \mathcal{M}^2 \otimes \mathcal{L}^{-1}$$

induced by ρ and the two different ways of mapping $\mathcal{M}$ into $\mathcal{M}^2 \otimes \mathcal{L}^{-1}$ and one obtains a sequence

$$0 \rightarrow \mathcal{L} \xrightarrow{\rho} \mathcal{M} \xrightarrow{\Delta} \mathcal{M}^2 \otimes \mathcal{L}^{-1}$$

where $\Delta = \varepsilon_1 - \varepsilon_2$. One checks by local arguments that this sequence is exact. In our situation $\mathcal{L} = \mathcal{F}_-^{\otimes d}$, $\mathcal{M} = \bigoplus \bigotimes p_i^* L^{\otimes d_i}$.

In a similar way as we produced the homomorphism ρ we can produce in our situation an embedding $\tau : \mathcal{M}^{\otimes 2} \otimes \mathcal{L}^{-1} \to \mathcal{M}^{\otimes 3}$ so that we obtain an exact sequence of coherent sheaves on X^m:

$$0 \to \mathcal{L} \xrightarrow{\rho} \mathcal{M} \xrightarrow{\delta} \mathcal{M}^{\otimes 3} .$$

By taking global sections one gets an exact sequence of K-vector spaces

$$0 \to \Gamma(X^m, \mathcal{F}_-^{\otimes d}) \to \Gamma(X^m, \otimes p_i^* L^{\otimes d_i})^{\ell} \to \Gamma(X^m, \otimes p_i^* L^{\otimes 3d_i})^{\ell^3} .$$

One next extends all these objects from Spec K to the base scheme Spec $\mathcal{O}_K$ and in this way endows the K-vector spaces above with an integral structure so that the above exact sequence of K-vector spaces induces an exact sequence of lattices over $\mathcal{O}_K$.

In the sequel we shall simplify our discussion on the geometry of numbers by taking $K = \mathbb{Q}$. The general case can be treated in a similar way as in section 5 by adelic methods. Let U, V, W be vector spaces of dimension a, b, c over the rationals and L, M, N lattices in U, V, W. Suppose that

$$0 \to L \xrightarrow{i} M \xrightarrow{\alpha} N$$

is an exact sequence of lattices and assume that V and W are euclidean vector spaces with euclidean form ε_V and ε_W. The corresponding norms are given by $|| \; ||_V$ and $|| \; ||_W$ and we let $\varepsilon_U = i^* \varepsilon_V$ be the form induced by ε_V on U with norm $|| \; ||_U = i^* || \; ||_V$. We normalize in such a way that V/M and W/N have volume 1. Finally let $\lambda(M)$ be the infimum of the maximum of the norms of basis vectors taken over all possible choices of a basis for M and $||\alpha||$ the norm of α. Then the following theorem replaces the arithmetic Riemann-Roch.

9.1 Theorem (Faltings, [F2]). *For* $0 \le i \le a-1$ *we have*

$$\lambda_{i+1}(L)^{a-i} \le 2^a b! \left(\frac{||\alpha||}{\lambda_1(N)} \right)^{b-a} \lambda_1(M)^{-i} \lambda(M)^b .$$

If we know upper bounds for the norms of α and a set of generators of M and lower bounds for the first successive minimum of N and M then the theorem gives upper bounds for the successive minima of L. If we know an upper bound for $\lambda_{i+1}(L)$ then we can find an element of L which is not contained in a given subspace U' of U of dimension $\le i$. In other words one can find elements of small norm which do not satisfy certain linear conditions.

This theory is applied to the above sequence of global sections. It was constructed in such a way that one obtains the required upper and lower bounds easily. In addition one needs to compute the dimensions of the second and third term in the sequence which can be obtained by the classical Riemann-Roch theorem. So instead of making difficult calculations for the first term it suffices to deal with the much more accessible spaces appearing in the second and third place.

The next point in our discussion is the intersection theory in $\mathbb{P}^n_{\mathbb{Z}}$ and the definition of the height of a cycle of pure codimension p in $P = \mathbb{P}^n_{\mathbb{Z}} \times \operatorname{Spec} \mathcal{O}_K$. On $\mathbb{P}^n$ the tautological line bundle $\mathcal{O}(-1)$ is defined. It is the subbundle of the trivial bundle on $\mathbb{P}^n$ of rank $n+1$ whose fibre over a point is the line defined by this point in the fibre of the trivial bundle over this point. Thus, if $\zeta_0, \ldots, \zeta_n$ are homogeneous coordinates in $\mathbb{P}^n$, $U_0 \subset \mathbb{P}^n$ the affine open set given by $\zeta_0 \neq 0$ and $z_j = \zeta_j/\zeta_0, j = 1, \ldots, n$, then we obtain a section s of $\mathcal{O}(-1)$ over U_0 which is given by

$$s(z_1, \ldots, z_n) = (1, z_1, \ldots, z_n) \in \mathbb{C}^{n+1} .$$

As a metric on $\mathcal{O}(-1)$ we can take

$$h^*(s, s) = 1 + |z_1|^2 + \ldots + |z_n|^2$$

and its curvature form becomes

$$\rho^* = -\sum_{i,j} \frac{\partial^2}{\partial z_i \partial \bar{z}_j} \log\left(1 + |z_1|^2 + \ldots + |z_n|^2\right) dz_i \wedge d\bar{z}_j \quad .$$

The curvature form ρ^* defines a Kähler metric on $\mathbb{P}^n(\mathbb{C})$ given by

$$\omega_0 = \sum_{i,j} \frac{\partial^2}{\partial z_i \partial \bar{z}_j} \log\left(1 + |z_1|^2 + \ldots + |z_n|^2\right) dz_i \otimes d\bar{z}_j \quad .$$

In this way we get an Arakelov variety P_{Ar} and the Arakelov Chowgroup $\mathbf{CH}_{\mathrm{Ar}}(P_{\mathrm{Ar}})$. To every irreducible subvariety $X \subseteq P$ of codimension p we define a class $(X, g_X) \in CH^p_{\mathrm{Ar}}(P_{\mathrm{Ar}})$ as follows. If X lies in a fibre over a finite place v of K then we put $g_X = 0$. Otherwise we put $\rho = -\rho^*$ and let

$$\rho_X = \deg X_K \cdot \rho^p$$

be the harmonic (p,p)-form on $P_\infty = P \otimes \mathbb{C}$ representing X. The form g_X is defined uniquely up to $\partial, \bar{\partial}$ by the conditions that

$$\frac{1}{\pi i} \bar{\partial} \partial g_X = \delta_X - \rho_X$$

and

$$\int g_X \cdot \omega = 0$$

for all harmonic forms ω. Also the first arithmetic Chern class $c_1(\mathcal{O}(1),h)_{\mathrm{Ar}}$ is defined where h is the metric on $\mathcal{O}(1)$ dual to h^*. The absolute height of X is then defined as

$$h(X) = \frac{1}{[K:\mathbb{Q}]}(X, g_X)\cdot(c_1(\mathcal{O}(1),h)_{\mathrm{Ar}})^{n+1-p}$$

where we have taken the intersection as in section 6. This can be extended by linearity to cycles of pure codimension p. One basic property of the height function is that it is non-negative for effective cycles of pure codimension p. As a consequence one shows that for X irreducible and flat over $\mathcal{O}_K$ and for a nontrivial section $f \in \Gamma(X, \mathcal{O}(d))$ one has

$$\frac{1}{2}\sum_{\sigma}\sup\log h_\sigma(f,f) \geq -[K:\mathbb{Q}]h(X)\cdot(\deg X)^{-1}\,.$$

Here h_σ is the restriction of the hermitian form h to $X_\sigma = X\otimes_\sigma\mathbb{C}$ and the supremum is taken over the manifold X_σ. Since X_σ is compact we could use the L^2-norm instead of the supremum norm. Both norms compare up to a factor of the form c^d for some positive constant c.

The last point we want to discuss is the product theorem. Let $P = \mathbb{P}^{n_1}\times\ldots\times\mathbb{P}^{n_m}$ be a multiprojective space over a field of characteristic zero, $d_1,\ldots,d_m$ positive integers and $L = \mathcal{O}(d_1,\ldots,d_m) = \otimes p_i^*\mathcal{O}(d_i)$ where $p_i : P \to \mathbb{P}^{n_i}$ denotes the i-th projection for $i = 1,\ldots,m$. As above we obtain on L a hermitian metric h so that (L,h) becomes a hermitian line bundle. The space P can be covered by affine open sets of the form $U = \mathbb{A}^{n_1}\times\ldots\times\mathbb{A}^{n_m}$. Over U the line bundle L becomes trivial so that a section f of L can be regarded over U as a function on U. If $x \in P$ is a point there exists such an open set U which contains x. We define then the index $i(x;f)$ of a section f of L at x as the maximum over all rational numbers σ such that for all non-negative integers $j_1,\ldots,j_m$ with $j_1/d_1+\ldots+j_m/d_m < \sigma$ and all differential operators D_i induced by a differential operator of degree at most j_i on $\mathbb{A}^{n_i}$ via the projection $p_i : U \to \mathbb{A}^{n_i}$ we have

$$D_1\ldots D_m f \in m_x$$

where m_x is the maximal ideal in the local ring at x. The index $i(x;f)$ is finite unless f is identically zero and does not depend on the choice of U. We fix a section $f \neq 0$ of L. Then f defines a stratification of P as follows. For a real number σ let Z_σ be the subset of P where $i(x;f) \geq \sigma$. Then $Z_\sigma \subseteq Z_\tau$ if $\sigma \geq \tau$. It is clear that Z_σ is defined locally by polynomials and is therefore a scheme.

Product Theorem (Faltings, [F2]). *For any given positive real number ε there exist positive real numbers r and c with the following properties. If $d_i/d_{i+1} \geq r$, $1 \leq i \leq m-1$, and if Z is an irreducible component of Z_σ and $Z_{\sigma+\varepsilon}$ then*

(i) $Z = Z_1 \times \ldots \times Z_m$, $Z_i \subseteq \mathbb{P}^{n_i}$ *closed*, $i = 1, \ldots, m$,

(ii) $\deg Z_i \leq c$.

The constants r and c can be computed effectively in terms of $\varepsilon, n_1, \ldots, n_m$. If K is a number field as usual and all objects are defined over K then in addition one can get bounds for the heights of the varieties Z_i. This theorem is then used in order to prove the finiteness theorem by induction.

10 An elementary proof of Mordell's conjecture

In this last section we discuss Bombieri's elementary approach to Mordell's conjecture. His version is a combination of Vojta's proof and Faltings' second proof discussed in section 9. The main modifications are the replacement of Arakelov theory by the elementary height theory introduced by Weil, and also the arithmetic Riemann-Roch theorem which was used in Vojta's work is replaced by an argument involving only Siegel's Lemma in combination with the classical geometric theorem of Riemann-Roch for algebraic surfaces which are even products of two curves. This technique was inspired by Faltings' proof of Lang's conjecture where he had introduced such a type of argument in a more general and more sophisticated way. Finally Bombieri does not use Arakelov varieties, i.e. varieties over the spectrum of the ring of integers of a number field. Instead it turns out that it suffices to work over the spectrum of a number field except that he has to use a modification of Eisenstein's theorem on the coefficients of Taylor expansions of algebraic functions. This is used in order to control denominators.

As usual let C be a non-singular curve of genus $g > 1$ defined over a number field K. We fix a point P in $C(K)$ and let $p_i : C \times C \to C$ be the projections onto the i-th factor for $i = 1, 2$. By Δ we denote as in section 3 the diagonal in $C \times C$ and we put $F_i = p_i^{-1}(P)$, $i = 1, 2$. Then as in the discussion of Vojta's proof we define $\Delta' = \Delta - F_1 - F_2$. Further we fix sufficiently large and divisible integers d, d_1 and d_2 with $gd^2 < d_1 d_2 < g^2 d^2$ and put

$$V = d_1 F_1 + d_2 F_2 + d\Delta' \ .$$

The precise choice of d, d_1 and d_2 will be made in the final step.

Next we express V as a difference of integral multiples of two very ample divisors which enables one to write sections of the line bundle $\mathcal{O}(V)$ locally as quotients of homogeneous polynomials and this is one of the main innovations for obtaining the elementary proof.

In order to get the decomposition of V one chooses a sufficiently large positive integer s such that

$$B = sF_1 + sF_2 - \Delta'$$

is very ample on $C \times C$ and therefore the pull-back of $\mathcal{O}(1)$ on $\mathbb{P}^n$ for some n. Similarly we fix a positive integer N such that $NF_1 + NF_2$ is the pull-back of $\mathcal{O}(1,1)$ on $\mathbb{P}^m \times \mathbb{P}^m$ for some m. Then we define for integers $\delta_i, i = 1,2$, the divisor

$$D(\delta_1, \delta_2) = \delta_1 NF_1 + \delta_2 NF_2 \ .$$

The integers δ_i are related to the parameters d_1, d_2, d and s by the equations

$$N\delta_i = d_i + sd \ , \quad i = 1,2 \ .$$

Clearly we obtain

$$V = D(\delta_1, \delta_2) - dB$$

and this is the representation we were looking for. If $y_0, \ldots, y_n$ generate the space of sections of $\mathcal{O}(B)$, then a section of $\mathcal{O}(dB)$ is a homogeneous polynomial of degree d in these generators. Similarly a section of $\mathcal{O}(D(\delta_1, \delta_2))$ can be written as a bihomogeneous polynomial of bidegree (δ_1, δ_2) in variables $x_0, \ldots, x_m$ and $x'_0, \ldots, x'_m$.

We shall now introduce the height of a section s of $\mathcal{O}(V)$. Clearly $y_i^d s$ is a section of $\mathcal{O}(D(\delta_1, \delta_2))$ and can be written as

$$y_i^d s = F_i(x, x') \ , \quad i = 0, \ldots, n \ ,$$

with a bihomogeneous polynomial F_i of bidegree (δ_1, δ_2). Its height $h(F_i)$ is defined in the usual way as $h(F_i) = \sum h_v(F_i)$ where $h_v(F_i)$ is the maximum of the logarithms of the v-adic absolute values of the coefficients of F_i. The sum is taken over all valuations of K. Then we put

$$h(s) := \max_i h(F_i) \ .$$

From this one sees that calculations with sections can be reduced to calculations with homogeneous polynomials and this is the key for the elementary approach.

We had fixed above a point P on $C(K)$ and this leads to an embedding j_P of C into its Jacobian $A = \mathcal{J}(C)$ given by

$$j_P : C \rightarrow A \ , \quad Q \mapsto \text{class}\,(Q - P) \ .$$

From this one gets the Riemann theta divisor θ. Its associated Néron-Tate height $\hat{h}_\theta$ defines a symmetric bilinear form

$$\langle x, y \rangle := \hat{h}_\theta(x + y) - \hat{h}_\theta(x) - \hat{h}_\theta(y) \ .$$

The right hand side of this equation is essentially the height attached to the Poincaré divisor corresponding to θ evaluated at (x, y) which again is related to the height given by the divisor Δ'. Using this remark it is easy to see that one gets an upper bound

$$h_V(z_1, z_2) \leq \frac{d_1}{2g}|z_1|^2 + \frac{d_2}{2g}|z_2|^2 - d\langle z_1, z_2\rangle + E_1$$

where z_1, z_2 are in $C(K)$, $|z_i|^2 = \langle z_i, z_i\rangle$ and E_1 is a lower order term with respect to $d, d_1, d_2, |z_1|, |z_2|$ being large.

If one can find a section $s \neq 0$ which does not vanish at (z_1, z_2) then a Liouville type estimate yields a lower bound for the left hand side. However one does not know in general a priori whether a section does vanish at a given point or not. This is precisely the situation where one needs multiplicity estimates. That is, one shows that at least some small "derivative of s" has the property not to vanish at the given point. In our context the classical Roth Lemma is precisely the tool one uses with success. If s is such a section then locally one can identify s with a function in two variables: Let D be the unit disc in $\mathbb{C}$ with the origin as center and $j : D \times D \to C \times C$ an immersion with $j(0,0) = (z_1, z_2)$. Then s induces a section σ of the trivial bundle $j^*\mathcal{O}(V)$ over $D \times D$ which can be identified with a function in two complex variables t_1, t_2. Then there exists a smallest value i such that

$$\left(\frac{\partial}{\partial t_1}\right)^{i_1} \left(\frac{\partial}{\partial t_2}\right)^{i_2} \sigma(0,0) \neq 0$$

for some pair (i_1, i_2) with $i_1 + i_2 = i$. As a consequence one gets after some calculation the lower bound

$$h_V(z_1, z_2) \geq -h(s) - c_1\left(i_1|z_1|^2 + i_2|z_2|^2\right) - E_2$$

where $E_2 \geq 0$ is again a lower order term. Here the modified Eisenstein estimate is used.

The next step is the elimination of the unknown section s. This is done with the help of Siegel's Lemma using the remark following the definition of the height of a section. It involves some condition on d_1, d_2 and d. Fix a real γ with $0 < \gamma < 1$ and restrict d_1, d_2 and d to the region

$$\frac{1+\gamma}{g} d_1 d_2 > d^2 .$$

Then there exists a section s such that

$$h(s) \leq c_2(d_1 + d_2)/\gamma + E_3$$

where again E_3 is a lower order term. As already mentioned the quantities i_1 and i_2 appearing in the lower bound for h_V are estimated from above with Roth's Lemma by

$$\frac{i_1}{d_1} + \frac{i_2}{d_2} = E_4$$

where E_4 is another lower order term with respect to the leading terms above. Such an estimate holds only provided we further restrict d_1 and d_2 to $d_1 \geq c_3 d_2$. Now we can compare the upper and lower bound for $h_V(z_1, z_2)$ and get the inequality

$$\frac{\langle z_1, z_2 \rangle}{|z_1|\,|z_2|} \leq \frac{d_1 |z_1|^2 + d_2 |z_2|^2}{2g\, d |z_1|\,|z_2|} + E_5$$

where we have collected in E_5 all the lower order terms $E_1, \ldots, E_4$ divided by $|z_1|\,|z_2|$. At this point we make the choices for d, d_1, d_2, z_1 and z_2. One wants to make the right hand side as small as possible. Clearly a good choice is to let $d_1|z_1|^2$ be roughly equal to $d_2|z_2|^2$. This means that we should define

$$d_i = \frac{t}{|z_i|^2} + E_{5+i} \ , \quad i = 1, 2 \ ,$$

where t is a new parameter which tends to infinity and E_6, E_7 are another pair of lower order terms. Then we are forced to put

$$d^2 = \frac{1+\gamma}{g} d_1 d_2 + E_8$$

which gives

$$d = \sqrt{\frac{1+\gamma}{g}} \frac{t}{|z_1|\,|z_2|} + E_9 \ ,$$

for lower order terms E_8 and E_9. Putting this into the above inequality, dividing by t and letting t tend to infinity one gets first

$$\frac{\langle z_1, z_2 \rangle}{|z_1|\,|z_2|} \leq \sqrt{\frac{1+\gamma}{g}} + E_{10}$$

where E_{10} depends on $\frac{1}{t}$ and if t is sufficiently large and γ sufficiently small we finally get

$$\frac{\langle z_1, z_2 \rangle}{|z_1|\,|z_2|} \leq \frac{3}{4} \ .$$

This leads to the following Theorem:

10.1 Theorem ([Bol]). *There exists an effective constant $c > 0$ such that for every pair of points z_1, z_2 on $C(K)$ with $|z_1| > c$ and $|z_2| > c|z_1|$ we have*

$$\langle z_1, z_2 \rangle \leq \frac{3}{4} \cdot |z_1|\,|z_2| \ .$$

From this theorem it is easy to deduce the finiteness theorem of Faltings.

10.2 Theorem ([Bo1], see also [Bo2]). *The points of $C(K)$ either have bounded height at most c, or they belong to a set of cardinality at most*

$$7^{\rho}(1+\log c/\log 2) \;,$$

where ρ is the rank of $A(K)$ and c is the constant in Theorem 10.1.

For the proof one uses Mumford's Theorem [M] on the growth of the heights of rational points on curves and applies Theorem 10.1.

10.3 Remark. Just before this book went into print it was pointed out by Bombieri in a letter to the author that in 10.2 the factor $1+\log c/\log 2$ can be replaced by $12(1+\log \deg C)$ where $\deg C$ is the degree of C for a non-singular complete projective embedding. So one can get for example the bound $\deg C \leq 2g+1$. Also the constant c can be estimated by a constant of the form $c(g)\cdot h(C)$ where $h(C)$ is the logarithmic height of C in a bicanonical embedding and $c(g)$ depends only on the genus g of C.

11 ℓ-adic Representations Attached to Abelian Varieties

In this final section we give a report on some of the work of Serre on ℓ-adic representations attached to abelian varieties. This work was very much influenced by the work of Faltings on the Tate module which was one of the main subjects of this book. Serre's extensive project aims to exploit three different aspects of these representations:

(i) to give an effective criterion which permits to detect when two such representations are isomorphic;

(ii) the determination of the algebraic envelop of ℓ-adic Galois groups;

(iii) the variation of the ℓ-adic Galois groups with ℓ.

Unfortunately the only published references are Serre's reports [S5] and [S6]. Here the results are stated but no proofs are given.

Let K be an algebraic number field, $\bar{K}$ an algebraic closure of K, $G_K = \mathrm{Gal}(\bar{K}/K)$ the Galois group of $\bar{K}$ over K and A an abelian variety of dimension $g \geq 1$ over K. We let ℓ be a prime number and $T_\ell = T_\ell(A)$ the Tate module of A. This is a free $\mathbb{Z}$-module of rank $2g$. We put $V_\ell = T_\ell \otimes \mathbb{Q}_\ell$. The Galois group G_K acts on T_ℓ so that we obtain an ℓ-adic representation

$$\rho_\ell : G_K \rightarrow \mathrm{Aut}(T_\ell) \cong \mathrm{GL}_{2g}(\mathbb{Z}_\ell) \ .$$

The image $G_{K,\ell} = \rho_\ell(G_K)$ is a compact subgroup of $\mathrm{Aut}(T_\ell)$. The family $(\rho_\ell)_\ell$ of such representations induce a homomorphism

$$\rho : G_K \rightarrow \prod_\ell G_{K,\ell} \subseteq \prod_\ell \mathrm{Aut}(T_\ell) \ .$$

The image of ρ is described in the following result.

11.1 Theorem (Serre, [S6]). *If K is sufficiently large $\rho(G_K)$ is an open subgroup of $\prod_\ell G_{K,\ell}$.*

On the other hand it is known from a result of Bogomolov that $G_{K,\ell}$ contains an open subgroup of the group $\mathbb{Z}_\ell^*$ of homotheties. It was conjectured by S. Lang (see [L3, I,§6], for a discussion of this) that this subgroup is the full group when ℓ is large. In this direction we have the following theorem.

11.2 Theorem (Serre, [S6]). *There exists an integer $n \geq 1$ such that $\rho(G_K)$ contains all homotheties of $\hat{\mathbb{Z}}^* = \prod_\ell \mathbb{Z}_\ell$ which are n-th powers.*

As a consequence the index of $G_{K,\ell} \cap \mathbb{Z}_\ell^*$ in $\mathbb{Z}_\ell^*$ is bounded independently of ℓ. It is not known whether n is effectively computable.

We next study the size of $G_{K,\ell}$. For this we fix a polarization E on A. Then we get on each T_ℓ an alternating form

$$E_\ell : T_\ell \times T_\ell \rightarrow \mathbb{Z}_\ell$$

which is a pairing of G_K-modules. The group $G_{K,\ell}$ is then contained in the group $\mathrm{GSp}(T_\ell, E_\ell)$ of symplectic similitudes with respect to E_ℓ.

11.3 Theorem (Serre, [S6]). *Suppose that the ring End(A) of $\bar{K}$-endomorphisms of A is $\mathbb{Z}$ and g is odd or 2. Then $G_{K,\ell}$ is open in $\mathrm{GSp}(T_\ell, E_\ell)$ for each ℓ and equal to $\mathrm{GSp}(T_\ell, E_\ell)$, if ℓ is sufficiently large.*

Again it is an interesting question to give an effective bound ℓ_0 such that for $\ell > \ell_0$ we have $G_{K,\ell} = \mathrm{GSp}(T_\ell, E_\ell)$. For results in this direction see section 2.

References

[BL] J.M. Bismut, G. Lebeau, Complex Immersions and Quillen Metrics, CRAS **309**(I) (1989), 487-491.

[BV] J.M. Bismut, E. Vasserot, The Asymptotic of the Ray-Singer Analytic Torsion Associated with High Powers of a Positive Line Bundle, Comm. Math. Phys. **125** (1989), 355-367.

[Bo1] E. Bombieri, The Mordell Conjecture Revisited, Annali Scuola Normale Sup. Pisa, Cl. Sci., S. IV, **17** (1990), 615-640.

[Bo2] E. Bombieri, Errata-Corrige The Mordell Conjecture Revisited, Annali Scuola Normale Sup. Pisa, Cl. Sci., S. IV, **18** (1991), 473.

[Bo-Va] E. Bombieri, J. Vaaler, On Siegel's Lemma, Invent. Math. **73** (1983), 11-32.

[Ch] D.V. Chudnovsky, G.V. Chudnovsky, Padé approximations and diophantine geometry, Proc. Natl. Acad. Sci. USA, **82** (1985), 2212-2216.

[F1] G. Faltings, Lectures on the Arithmetic Riemann-Roch Theorem, to appear in Annals of Math. Studies, Princeton University Press.

[F2] G. Faltings, Diophantine approximation on abelian varieties, Annals of Math. **129** (1991), 549-576.

[F3] G. Faltings, The general case of S. Lang's conjecture, Preprint (1991).

[F-Ch] G. Faltings, C.L. Chai, Degeneration of Abelian Varieties, Ergeb. d. Math. 3. Folge, Band **22** (1990), Springer-Verlag.

[GS1] H. Gillet, C. Soulé, Arithmetic intersection theory, preprint 1988, see also CRAS **299**(I) (1984), 563-566.

[GS2] H. Gillet, C. Soulé, Characteristic Classes for Algebraic Vector Bundles with Hermitian Metric, Annals Math. **131** (1990), 163-203, 205-238.

[GS3] H. Gillet, C. Soulé, Analytic Torsion and the Arithmetic Todd genus, Topology **30** (1991), 21-54.

[GS4] H. Gillet, C. Soulé, Amplitude arithmétique, CRAS **307** (1988), 887-890.

[K] S. Kawamata, On Bloch's Conjecture, Invent. Math. **57** (1980), 97-100.

[L1] S. Lang, Introduction to Arakelov theory, Springer 1988.

[L2] S. Lang, Integral points on curves, Publ. Math. IHES **6** (1960), 27-43.

[L3] S. Lang, Number Theory III, Encycl. Math. Sc. **60** (1991), Springer-Verlag.

[MW1] D.W. Masser, G. Wüstholz, Estimating isogenies on elliptic curves, Invent. Math. 100 (1989), 1-24.

[MW2] D.W. Masser, G. Wüstholz, Some effective estimates for elliptic curves, in: "Arithmetic of complex manifolds", SLN 1399 (1989), 103-109.

[MW3] D.W. Masser, G. Wüstholz, Periods and minimal abelian subvarieties, to appear in Annals of Math.

[MW4] D.W. Masser, G. Wüstholz, Isogeny estimates for abelian varieties, and finiteness theorems, to appear in Annals of Math.

[MW5] D.W. Masser, G. Wüstholz, Isogenies and endomorphisms of abelian varieties, manuscript.

[MW6] D.W. Masser, G. Wüstholz, A note on polarizations of abelian varieties, manuscript.

[MW7] D.W. Masser, G. Wüstholz, Galois properties of division fields of elliptic curves, manuscript.

[M] D. Mumford, A remark on Mordell's conjecture, Am. J. Math. **87** (1965), 1007-1016.

[Ra] M. Raynaud, Around the Mordell conjecture for function fields and a conjecture of Serge Lang, in: "Algebraic geometry", SLN 1016 (1983), 1-19.

[Ro] K.F. Roth, Rational approximations to algebraic numbers, Mathematika 2, **1955**, 1-20.

[S1] J.-P. Serre, Abelian ℓ-adic representations and elliptic curves, Benjamin 1968.

[S2] J.-P. Serre, Propriétés galoisiennes des points d'ordre fini des courbes elliptiques, Invent. Math. **15** (1972), 259-331.

[S3] J.-P. Serre, Quelques applications du théorème du densité de Chebotarev, Publ. Math. I.H.E.S. 54 (1981), 123-201.

[S4] J.-P. Serre, Local fields, GTM **67** (1979), Springer-Verlag.

[S5] J.-P. Serre, Résumé des Cours et Travaux, in "Annuaire du Collège de France", 1984-1985, 85-91.

[S6] J.-P. Serre, Résumé des Cours et Travaux, in "Annuaire du Collège de France", 1985-1986, 95-100.

[V1] P. Vojta, Mordell's conjecture over function fields, Invent. Math. **98** (1989), 115-138.

[V2] P. Vojta, Siegel's theorem in the compact case, Annals of Math. **133** (1991), 509-548.

[V3] P. Vojta, Dyson's Lemma for products of two curves of arbitrary genus, Invent. Math. **98** (1989), 107-113.

[V4] P. Vojta, Integral points on subvarieties of semiabelian varieties, preprint (1991).

[We] A. Weil, Basic number theory, Grund. math. Wiss. (1973), Springer-Verlag.

[W1] G. Wüstholz, Algebraische Punkte auf analytischen Untergruppen algebraischer Gruppen, Annals of Math. **129** (1989), 501-517.

[W2] G. Wüstholz, Multiplicity estimates on group varieties, Annals of Math. **129** (1989), 471-500.

Aspects of Mathematics

Edited by Klas Diederich

*A publication of the Max-Planck-Institut für Mathematik, Bonn

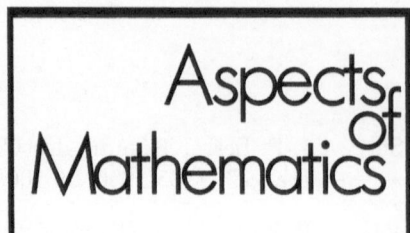